Jean Massart, Emile Vandervelde, Jean Demoor

Evolution by Atrophy in Biology and Sociology

Jean Massart, Emile Vandervelde, Jean Demoor

Evolution by Atrophy in Biology and Sociology

ISBN/EAN: 9783337215651

Printed in Europe, USA. Canada, Australia, Japan

Cover: Foto ©berggeist007 / pixelio.de

More available books at **www.hansebooks.com**

EVOLUTION BY ATROPHY

IN BIOLOGY AND SOCIOLOGY

BY

JEAN DEMOOR
AGRÉGÉ OF THE FREE UNIVERSITY
OF BRUSSELS

JEAN MASSART
CHARGÉ DE COURS OF THE FREE
UNIVERSITY OF BRUSSELS

ÉMILE VANDERVELDE
PROFESSOR AT THE INSTITUTE OF HAUTES ÉTUDES OF BRUSSELS

TRANSLATED BY

Mrs CHALMERS MITCHELL

LONDON
KEGAN PAUL, TRENCH, TRÜBNER & CO., Ltd.
PATERNOSTER HOUSE, CHARING CROSS ROAD
1899

CONTENTS

	PAGE
PREFACE	5

INTRODUCTION

1. Societies and organisms. 2. Individuals, colonies and societies. 3. Communities and societies. 4. Distinctive characters of societies of which the members are united by social contract 7

BOOK I

UNIVERSALITY OF DEGENERATIVE EVOLUTION

PART I. Degeneration in the development of institutions and organs 21

Chapter I. In the evolution of organs all modification is necessarily attended by degeneration . . 22

§ 1. Preliminary considerations . . . 22
    (1) Original formation of organs in ancestral forms . 23
    (2) Development of organs in the individual . 23
    (3) Philogenetic evolution of function . 25
    (4) Individual adaptation . . . 26
    Section I. Transformation of organs of animals . 30

§ 2. Transformation of homodynamic organs in the individual (metameric appendages) . . 30
    The cray-fish 31

§ 3. Transformation of homologous organs in individuals of different species (limbs) . . . 41
    Limbs adapted to an aquatic life (*Ceratodus, Orthacanthus, Protopterus amphibius, Protopterus annectens, Lepidosiren* 43
    Limbs adapted to a terrestrial life . . 43

CONTENTS

	PAGE
1. Adaptation to walking on two legs (man, birds). 2. Adaptation to leaping (*Dipus ægyptius*, kangaroo, *Tarsius spectrum*, *Rana esculenta*). 3. Adaptation to running (horse, ruminants). 4. Adaptation to flight (birds, Pterosaurians, bats). 5. Adaptation to arboreal life (*Arctocebus calabarensis*, *Chamæleo*). 6. Adaptation to swimming (Cetaceans, Sirenia). 7. Adaptation to burrowing (*Talpa europæa*, *Heterocephalus*) . . .	46

Section II. Modification of the organs of plants . 68

§ 4. Modification of homodynamic organs in the individual (basilar and apical leaves) . . . 68

1. *Rosa rugosa*. 2. *Serratula centauroides*. 3. *Sagittaria sagittifolia*. 4. *Lathyra Aphaca*. 5. *Nymphaea dentata* 70

§ 5. Modification of organs which are homologous in individuals of different species (foliage leaves) . 78

1. Adaptation to climbing (*Cobaea scandens*, *Vicia pyrenaica*, *Cucumis sativus*). 2. Adaptation to carnivorous nutrition (*Utricularia*, *Nepenthes*, *Drosera*). 3. Adaptation to an aquatic life (*Sagittaria*, *Nymphaea*, *Vallisneria*, *Potamogeton*, *Ranunculus*, *Ouvirandra fenestralis*). 4. Adaptation to defence against ants (*Acacia sphaerocephala*). 5. Adaptation to drought (*Sempervivum*). 6. Adaptation to defence against herbivorous animals (*Caragana*, *Ilex*, *Mamillaria*, etc.) . . 78

Chapter II. In the evolution of institutions all modification is necessarily accompanied by degeneration . . 90

§ 1. Modifications of similar institutions in the same society 91

(1) The communal budgets of Belgium . . 92
(2) Budget of the States of the German Empire . 95
(3) The budgets of Germany, France, and England . 97

§ 2. Modification of similar institutions in different social groups (the development of landed property) . 98

1. Family property (Montenegro). 2. Village property (Russia). 3. Feudal property (England).

CONTENTS

	PAGE
4. Public or collective property (Switzerland). 5. Corporative property (Belgium). 6. Private property (Switzerland). 7. Summary	100
PART II. Degeneration in the evolution of organisms and societies	115
Chapter I. All organisms exhibit rudimentary organs	115
Section I. Rudimentary organs of animals	117
§ 1. Rudimentary organs in man	117
1. Integumentary system. 2. Skeleton. 3. Muscular system. 4. Nervous system. 5. Digestive system. 6. Vascular system. 7. Sense organs. 8. Genito-urinary system	117
§ 2. Rudimentary organs in various groups	121
1. Cœlenterates. 2. Worms. 3. Echinoderms. 4. Mollusca. 5. Arthropodes. 6. Vertebrates	121
Section II. Rudimentary organs in plants	145
§ 3. Rudimentary organs in various groups of plants	145
1. Algæ. 2. Mushrooms. 3. Bryophyta. 4. Pteridophyta. 5. Phanerogams	145
§ 4. Reduced organs in the vegetative apparatus of the Phanerogams	149
Chapter II. Survivals exist in all kinds of societies	151
§ 1. Instances of survival in various groups	153
(1) Instances of survival in the most modern social groups (the United States)	155
(2) Instances of survival in less civilized social groups (Veddahs, Fuegoes, Australian tribes	156
§ 2. Survivals of ancient forms of marriage, and of the family in Modern Europe	161
1. *Forms of marriage.*—(1) Marriage by capture. (2) Marriage by purchase. (3) Marriage by consent of both parties (marriage by simple consent, marriage *in facie Ecclesiæ*)	161
2. *The Family System.*—(1) Matriarchy. (2) Patriarchy	167
PART III. Summary and conclusions	170

CONTENTS

BOOK II

THE PATH OF DEGENERATIVE EVOLUTION

	PAGE
PART I. The supposed law that degeneration retraces the steps of progress	175
Chapter I. The path of degeneration in biology	178
Section I. The path of degeneration in animals	179
1. Morphology and embryology; the law of recapitulation. 2. Degeneration of the third eye in lizards. 3. Degeneration of the organs of sight in deep-sea Crustacea. 4. Atrophy of the branchial vessels in man	179
Section II. The path of degeneration in plants.	192
1. Rarity of cases of recapitulation in the organogeny of leaves (Vicia, *Acacia* with phyllodes). 2. Organogeny of flowers (*Brassica oleracea* var. *Botrytis*). 3. Progressive degeneration of the prothallus in phanerogams	192
Chapter II. The path of degeneration in sociology	205
§ 1. Investigation of facts	205
1. Tithings, hundreds and counties in England. 2. Order of elimination of various racial elements in a country. 3. The degenerative evolution of political organizations. 4. Degeneration in monetary systems. 5. Degenerative adaptation in colonial legislation. 6. Degenerative evolution of the corporations of Western Flanders	207
§ 2. A criticism of the supposed inverse path of degeneration	217
PART II. The irreversibility of degenerative evolution.	221
Chapter I. Do institutions or organs which have disappeared reappear?	222
Section I. Disappeared organs	222
1. Plants	222
2. Animals. Teratology of the horse, Hypertrichosis, etc. Swimming limbs in Stomatopoda and Decapoda Macroura	223
Section II. Disappeared institutions	227

CONTENTS

	PAGE
(1) Apparent revival of bygone institutions	227
(2) Apparent disappearance of institutions	229
(3) Instances of convergence	230

Chapter II. Can rudimentary institutions or organs reassume their primitive functions? 232

   Section I. Rudimentary organs 232

      1. *Animals.* (1) Muscles of the ear in man. (2) The abdomen and appendages in deep-sea hermit crabs . . 232

      2. *Plants.* (1) Hermaphrodite flowers in *Melandryum.* (2) Branches of *Colletia cruciata, Crataegus, Vicia Faba, Sempervivum, Veronica,* etc. . . . 233

   Section II. Rudimentary institutions . . . 237

      (1) Truck system and clearing-house. (2) Corporations and syndicates. (3) Archaic collectivism and modern collectivism. (4) The survival of elective sovereignty in England 239

Chapter III. Can rudimentary organs or institutions redevelop and assume new functions? . . . 242

   Section I. Rudimentary organs . . . 243

      1. *Animals.* Respiratory organs in *Birgus latro;* Mesonephric spaces in the higher vertebrates . 243

      2. *Plants.* Staminodes of *Pentstemon* . . 244

   Section II. Rudimentary institutions . . . 245

      Levirat 245

PART III. Summary and conclusions . . . 247

BOOK III
CAUSES OF DEGENERATIVE EVOLUTION

PART I. Atrophy of organs and institutions . . 251

   Section I. The factors of atrophy . . . 251

      (1) Biology (accidental, individual, normal and specific atrophy). (2) Sociology (accidental and normal atrophy) 251

   Section II. Causes producing atrophy . . . 260

   Chapter I. Atrophy of organs 260

      § 1. Atrophy from lack of space . . . 261

CONTENTS

	PAGE
(1) Development of the teeth. (2) Atrophy of the superior glume (*Lolium*). (3) Degeneration of Paleæ (composite flowers) and of stamens (Scrophulariaceæ)	261
§ 2. Atrophy from lack of use	263
1. Functional inutility: (1) Etiolated plants and immobile limbs. (2) Epicotyl and primary leaf of *Nymphaea*. (3) Roots of *Utricularia*; cotyledons of parasitic plants (*Cuscuta, Orobanche*); leaves transformed to spines in *Phyllocactus crenatus*. (4) Atrophy of the branchial arches in mammals. (5) Atrophy of ventral fins (*Pediculati, Protopterus*). (6) Atrophy of muscles (*Cetacea, Sirenia*). (7) Atrophy of the tail in man. (8) Degeneration of the hyoid apparatus in man and birds	263
2. Transference of function: (1) Atrophy of the tail in *Batrachia anura*, and the larval gills in some insects. (2) Disappearance of limbs (Slow-worms, Amphisbæna, Snakes, Eels, and Sacculina). (3) Atrophy of the leaf (*acacia* with phyllodes, *Xylophylla*). (4) Atrophy of the protonema in mosses, and of the leaves in some xerophilus plants (*Muehlenbeckia platyclados, Genista, Spartium, Alhagi*). (5) Disappearance of the calyx. (6) Atrophy of roots (Pine, Beech, *Corallorhyza, Myrmechis, Tillandsia usneoides*), or of leaves and stems (*Tæniophyllum Zollingeri, Podostemaceæ*)	268
§ 3. Atrophy from lack of nutrition	274
1. Parasitic castration (*Melandryum album*). 2. Severe or prolonged compression of a limb. 3. Atrophy of the genitalia in neuter bees. 4. Atrophy of the superior flowers in *Carex*. 5. Atrophy of pistils and stamens (*Fritillaria persica, Viburnam tomentosum, Viburnam Opulus*)	274
§ 4. Atrophy without apparent cause	278
Atrophy of perianth (*Artemisia, Poterium*). Atrophy of the eyes (Myriopodes, *Cymothoë*). Correlative atrophy	278

CONTENTS

	PAGE
Chapter II. Atrophy of institutions	281
§ 1. Atrophy from lack of use	282
1. *Functional inutility:* (1) Offices of the port of Bruges. (2) The forest-courts of England	282
2. *Transference of function:* (1) Republican institutions under the Roman Empire. (2) Special jurisdiction in England	284
§ 2. Atrophy from lack of resource	287
1. Local administration at the close of the Roman Empire. 2. Degeneration of societies in general	287
PART II. Causes of the persistence of organs or institutions without function	292
Chapter I. Survival of organs	292
§ 1. Unfunctional organs that are not rudimentary	292
Absence of variation (flowers of *Ficaria ranunculoides, Lysimachia Nummularia, Elodea canadensis, stratiotes aloides,* cleistogamous flowers, eyes in the male *Machaerites*	292
§ 2. Unfunctional organs which persist as rudiments	295
1. Absence of variation. Insignificance of the rudimentary organ (stipules of *Tropaeolum majus*, accessory rudiments of enamel organs in the development of teeth	295
Chapter II. Survival of institutions	298
§ 1. Integral persistence of an institution	299
(1) Maintenance by compulsion (rotten boroughs in England; the States of Normandy and the Dauphiny in France). (2) Indirect usefulness (English monarchy). (3) Respect for tradition (royal prerogatives exercised by the prætor; institution of sheriffs in England)	299
§ 2. Survival of institutions in a reduced condition	306
1. Insignificance of the institution (jurisdiction at da Martinique; summons *quo warranto*; Diocese of Cambrai; marsh-land in Artois; tribute paid to Spain by France). Respect for traditions (survival of the old regime in England; instances of survival in law and religion	307
PART III. Summary and conclusions	317
General conclusions	320

PREFACE

THIS treatise, compiled in connection with a scheme for research work in general sociology, elaborated in June 1894, was presented to the Institute of Sociology at Brussels. In drawing up the programme of the Institute, the founder, M. Ernest Solvay, after having mentioned the questions which especially called for the investigation of his colleagues, added the following statement:—

"The Institute of Sociology will take part in the labours of the modern school of Sociology, the object of which is to ascertain the normal conditions under which societies exist, and the laws which govern their evolution. But the advances of Natural Science in this century have not yet been sufficiently assimilated by those Sciences most closely related to it, and it is from such assimilation that the most important additions to knowledge may be expected."

In stating that the results of Natural Science have not been sufficiently assimilated by Sociology, M. Solvay is only apparently at variance with those who rightly protest against exaggerated and premature comparisons between social organizations and animal or vegetable organisms.

The existence of such exaggerations, which have caused a reaction such as recently induced an eminent American economist to declare the bankruptcy of biological sociology, is perhaps due to the fact that, with a few distinguished exceptions, bio-sociological investigations have hitherto been conducted either by naturalists with a limited knowledge of social questions, or by sociologists whose training in biology was incomplete and superficial.

To prevent this danger, our researches in the same subject have been made separately from the social side, and from the biological side, and have now been co-ordinated and combined.

This work was commenced in May 1893, with the collaboration of our friend M. Dollo, the curator of the Natural History Museum at Brussels. In June 1894, however, M. Dollo's many occupations no longer permitted of his collaboration. The zoological part was therefore completed by M. Jean Demoor, to whom most of the facts quoted in the first book were given by M. Dollo, whose assistance we most gratefully acknowledge.

INTRODUCTION

EVER since the application of theories of evolution to social phenomena, there has been a constant interchange in terminology between biology and sociology; societies have been called organisms, and organisms societies of cells. There is an actual division of labour among the organs of a living body, while institutions have been called the organs, or parts of organs, of Society. The interchange of matter effected among the organs of an individual has been called a "physiological contract," while the circulation of money may be compared to the circulation of blood and lymph. Questions arise as to what extent such comparisons are legitimate, if they should be taken in any other than a metaphorical sense, and if it is possible to set a precise boundary between the provinces of biology and sociology.

Much has already been written on such problems as these, and no doubt much more will yet be written. We shall not attempt either to discuss or to solve them in these few introductory pages; they are merely touched upon here, and will only be alluded to when absolutely necessary for the careful investigation of facts bearing upon our work.

I. *Societies and organisms.*—The analogy existing, from the point of view of evolution, between biology and sociology, arises from the fact that the evolution of societies as well as that of organisms, is the result of the co-operation of two factors—similarity and adaptation.

In biology, the similarity between organisms springing from the same stock, is due to heredity, while adaptation is the result of individual variation.

In sociology, societies are the descendants of former societies, in that the new are modelled upon the old. Similarity is the result of imitation, while adaptation is the result of invention—*i.e.* of all improvement and innovation tending to make a new society different from that which preceded it.[1]

These fundamental analogies suffice in themselves to justify our collaboration, whatever may be the solution of the question—which is really only a question of terms—as to whether societies should be regarded as organisms, or organisms as particular kinds of societies.[2]

---

[1] Invention is frequently a combination of several imitations. When a society is formed, its characters are not necessarily borrowed from those societies from which it more or less directly proceeds. It may be modelled upon other social structures with which it had no hereditary connection.—V. Tarde, *Les Lois de l'imitation*.

[2] See *Les Sociétés animales*, p. 128 (Espinas). "Integration, or grouping together, is a universal law common to all organic or inorganic existence. Society, properly speaking, is only a complex and important instance of this universal law."

See *La Science Sociale*, p. 97 (Fouillée). "All purely physio-

In either case, it is certain that organisms and societies—used in the sociological acceptation of the term—exhibit some characters in common, and some distinctive characters.

The common characters may be accounted for by the co-operation which exists in both cases between the units of which they are composed (individuals, or cells).

The dissimilar characters are probably connected with what constitutes the essential difference between social aggregates and organic aggregates. With the former there is a physiological continuity between the composing units, while with the latter co-operation is entirely due to mental relations.

II. *Individuals, colonies and societies.*—Our view in this matter is considerably at variance with the current opinion. Many authors, and M. Espinas among them, regard colonies, whether animal or vegetable, as societies, even when the members of these colonies are connected by physiological bonds.[1]

logical characters of life, viz.: 1. Correlation of parts; 2. Relation between structure and function; 3. Division of all living parts into other living parts; 4. Spontaneity of movement; 5. Particulate existence; 6. Development and degeneration, *i.e.* evolution, are to be found in a greater extent in animal and human societies."

[1] See *Les Sociétés animales,* section 2, pp. 207 and foll.—(Espinas), *Cours de philosophie positive,* vol. iv. (A Comte). "These strange societies are to be found among the lower animals, an involuntary co-operation being the result of an unseverable organic union, which is either a mere adherence or actual continuity."

We shall justify the distinction we make by showing that colonies are entirely distinct from societies. The two are divergent branches which spring from the same source, a solitary individual. These divergent branches of organic life may be distinguished in the following manner: An individual may be either unicellular or multicellular. Every cell capable of sustaining life and reproducing its own kind is an individual.[1]

In the case of multicellular individuals there is an unbroken physiological continuity, while life lasts, between the cells of which they are composed. All these units spring from one primary cell (*i.e.* a fertilized egg) which sprang itself from two cells (male and female), proceeding usually from two separate individuals. From the moment that this union is effected, the continuity remains unbroken.[2] When such an individual reproduces

---

[1] We regard the cell as the unit of life, although, according to Altmann, the unit of life is a still simpler structure—the bioblast. He maintains that the bioblast, or living granule, is the ultimate element of the cell; it is born from a pre-existing granule; it lives, feeds, reproduces, and is, in fact, an organism (see *Die Elementarorganismen und ihre Beziehungen zu den Zellen*; Leipzig, 1890.—*Studien über die Zellen*; Leipzig, 1886.—*Zur Geschichte der Zelltheorien*; Leipzig, 1889 (Altmann). The existence, however, of the bioblast is a pure hypothesis, no definite proof having been established of its existence; and to assume that it is the simplest unit of life is to abandon oneself to pure imagination.

[2] Although this theory is in agreement with the observed facts of biology, it is necessary to point out the complexity of the ideas of individual and organism. Boveri (*Über die Befruchtungs*

asexually, giving rise to offspring which remain attached to it, the whole forms a colony. All the cells of this colony spring then from one single cell (a fertilized egg cell), and an interchange of nutriment goes on incessantly among these cells. This is what Spencer designates as a "physiological contract." Societies, on the other hand, consist of units springing from various sources, whose connection is merely mental. It is only in the communities which Tonnies calls "*Gemeinschaften*," in distinction from societies bound by mental agreement (*Gesellschaften*), that all the individuals spring from one and the same couple.

From our point of view, a colony must not be regarded as an intermediate condition between an individual and a society. No known society has passed through a colonial stage, and the members of a colony could not, on separating, become a society. Among the simplest aggregates, all the units are

und Entwicklungsfähigkeit kernloser Seeigel-Eier, und über die Möglichkeit ihrer Bastardirung. Arch. f. Entwicklungsmechanik der Organismen, Bd. ii., pp. 394-443), for instance, has shown that it is possible to fertilize the eggs of Echinoderms, from which the nuclei have been removed, by the spermatozoa of other species of Echinoderms. The egg is a typical cell, an organism, an individual. All its parts are essential to it, and are incapable of separate existence, at any rate for any length of time; close physiological bonds unite the component parts, yet it is possible to substitute for an essential part of one individual a part taken from another individual. It may therefore be concluded that the idea of an organism no longer necessarily implies the idea of continuous functional unity which one was formerly tempted to ascribe to it.

equivalent, and each retains all the functions of life (*Spirogyra*, Choanoflagellates, Hydra). The colony may disintegrate, but the cells or individuals set at liberty are capable of living alone, and do not form a society.[1]

As the organisation of a colony becomes more complicated, the units proceed to differentiate. Each one assumes certain functions in particular, and becomes less adapted to discharge others. If such a colony is dispersed, the individuals are incapable of maintaining a separate existence.[2]

---

[1] (*a*) A thread of *Spirogyra* consists of cells placed end to end, physiologically connected with one another. Under certain conditions, however, quite apart from the phenomenon of cellular reproduction, all these cells are capable of isolating themselves. The colony is thus transformed into a set of individuals capable of maintaining a separate existence, and no longer in connection with one another.

(*b*) Among the Choanoflagellates, some consist of free individuals, while in others all the individuals are united by a common stalk, and intercommunicate by protoplasmic threads in the stalk. When, for some cause, these individuals separate, they never form themselves into a society.

(*c*) A colony of Hydra is formed by the budding of a single individual. While nourishment is abundant, all the individuals of the colony retain their connection with each other, and may themselves give rise to buds. On the other hand, when food is scarce, the colony disintegrates and each individual lives a free life without entering into social relations with its neighbours.

[2] In some cases the terms individual and colony become extremely involved. Although one accept our view, it is difficult to rigorously apply our generalizations to such facts as the following: Certain male cephalopods, at the period of reproduction, separate from their bodies a specialized tentacle (hectocotyle) in which is stored the seminal fluid. This organ, set at liberty, swims in the

III. *Communities and societies.*—The bond existing between all the constituent parts of a society is not of the same nature as that which unites the members of a colony or the cells of an animal or plant. These physiological bonds are not compar-

sea, and after living there a certain time, enters the female cephalopod, to effect impregnation.

The individuality of this hectocotyle appears so obvious that for a long time it was regarded as a distinct species—some kind of worm. At first sight, then, the hectocotyle seems to be an individual. This, however, is not the case, as it reproduces, not a hectocotyle but a cephalopod.

Many Echinoderms, by a spontaneous act of protection, can separate their arms from their bodies. Such a separated arm may live, feed and slowly build up again a whole Echinoderm. Plainly, it would be impossible to indicate the precise point at which such an organ should be regarded as an individual. The vegetable kingdom abounds in analogous facts. A strawberry plant, for instance, gives off runners in the course of the Summer, which take root, and themselves become strawberry plants. So long as these young plants are insufficiently developed to maintain themselves, the mother-plant continues to supply them with nourishment. As soon, however, as the young shoot can dispense with this support, the runner atrophies, and the little plant begins a separate existence. By prematurely cutting the runner, the new plant may be compelled to live alone sooner than it naturally would have done.

With some other plants (*Phalangium viriparum*) the young shoots frequently retain their connection with the mother-plant, although quite capable of maintaining themselves. Under these circumstances it is obviously impossible to say if these plants represent colonies or free individuals. Speaking generally, it may be said that no precise line can be drawn between colonies and individuals. Many writers on the subject, and Perrier among the number, consider that every colony where there is a physical continuity among the members, should be regarded as an individual.

able, without forcing the analogy, although they have been so compared by some sociologists, to such means of communication between individuals and societies as exchange, traffic, roads, railways, telegrams and telephones. It is merely a matter of definition, and if societies are to be termed organisms, they should be distinguished as organisms by social contract, *organismes contractuels* (Fouillée).

This definition, however, only applies to those societies which owe their existence to a formal contract with definite objects in view, and not to ready-made communities consisting of individuals already united together without any preliminary contract. The latter is the case, for instance, in societies of ants or bees, and in human societies in those social groups in which the individuals are united by the bonds of consanguinity. The characters of such communities partly approach the characters of organic associations, but precisely as such natural communities approach societies by social contract, the differences between social groups and actual organisms become more marked. In the more complex forms of societies the results of the characters we have distinguished become most accentuated.

IV. *Distinctive characters of societies of which the members are united by social contract.*—(1) A cell cannot be part of two organisms or of two organs at the same time.¹ On the other hand, there is no reason why the members of one social community should not belong to other communities at the same time.

2. Speaking generally, the biological conception of an organism denotes a definite thing—a plant or an animal in itself, quite distinct from similar organisms. In sociology, however, there is no precise line between co-existing social communities. Are we to regard, for instance, the families, communes and cantons of a state as distinct organisms, or merely as organs? Does a free town such as Hamburg or Frankfort cease to be an organism when it loses its independence? Take the various Swiss cantons, which are now mere organs of the Helvetic Confederation, like the Provinces of Belgium, or the Departments of France, would they become organisms on the rupture of the Federal bond? On the other hand, with the growth of international treaties between the states of Eastern Europe, will their social individuality disappear, and will they come to be regarded as are the United States of America, as the organs of an organism in process of formation? These few examples suffice to show

¹ When two organs are united into one whole, the cell exercises two totally different functions. The liver, as we now know, is a double organ consisting of a bile-forming liver, and the glycogen-producing liver, two organs which are embryologically distinct. The cells of which the liver is composed are both bile-secreting and glycogen-forming organs.

that, so far as social matters are concerned, the conception of organisms is a pure convention. In the course of this treatise, we may therefore regard families, societies and nations as distinct organisms, or, with regard to their connection with other and vaster organizations, as organs of the latter.

3. The structure of an animal or plant depends upon the physical arrangements of its parts, and on the physiological links between those parts. The structure of a society depends upon the links of social contract existing among its members. We regard these as two very different things, and we cannot follow Tarde in pressing the analogy between them in the following way: "The length, breadth, and height of an organism are never very much out of proportion. With snakes and poplars the height or length preponderates; among flat fish the thickness is very small compared to the other dimensions, but in each instance the disproportion exhibited in extreme cases is not comparable to that shown by any social aggregate—such as China for instance, which is 3000 kilometres in length and breadth, and only one or two yards in average height, for the Chinese being a short race, build their edifices correspondingly low."[1]

4. Organic modifications are effected more slowly and with greater difficulty than are social modifica-

[1] *Les Monades et la Science sociale.* (Tarde. *Revue de Sociologie*, 1893, p. 169.)

tions.[1] The result is an important one from the point of view of method.

In biology, excepting in the case of individual adaptation of artificial selection, direct observation —the historical method, if we may so call it—is not available for the study of the origin and modification of organisms. Phylogeny, the science of organic kinship, resorts to other methods, and particularly to the comparative method in its various forms:—

(*a*) Morphology, the science of determining the phylogeny of organs by comparing them with the organs of other creatures belonging to the same systematic group.

(*b*) Palæontology, which determines the direct ancestors of living creatures.

(*c*) Embryology, which, so far as it is founded upon the principle of recapitulation, investigates the development of organs in the individual, and draws conclusions therefrom bearing upon its descent.

---

[1] Among animals there is a special factor which gives a stability to the specific characters not found elsewhere—this is reproduction. Specific characters being common to the whole line of descent, are very deeply enrooted in the organism. They are not easily modified by the influence of new environments, but maintain their likeness to one another in spite of external conditions. They are regulated by an internal force, notwithstanding the importunities to variation offered from outside. This force is heredity, and heredity accounts for the precise way in which specific characters may be defined. In society this internal force is wanting. (*Les Règles de la méthode sociologique.* Durckheim, Paris, F. Alcan, 1895.)

(d) Teratology, which compares normal with abnormal forms for the same purpose.

In sociology these various systems of comparison are only of a secondary value, owing to the great variability of social forms, while the historical method of investigation assumes a greater importance. In sociology, however, there are methods analogous to the comparative methods of biology:

(a) Archæology corresponds to palæontology.

(b) Social morphology, by comparison of series of institutions, makes up to a certain extent for the absence of direct observation of their origin and development. Thus, to use the phrase of Bagehot, by studying the customs and institutions of modern savages, the prehistoric living may be made to throw light upon the prehistoric dead. If an institution be found in full operation among savages, of which a vestige still exists among more civilized people, it may be assumed that the vestige was at some time fully functional among the latter. It must be borne in mind, however, that in many instances such vestiges are the result of imitation. Theodore Reinach has shown that this applies in the case of circumcision.[1]

This gravely weakens the conclusions drawn by Spencer from the survival of this custom among

---

[1] *De quelques faits relatifs à l'histoire de la circoncision chez les peuples de la Syrie* (Th. Reinach, *L'Anthropologie*, 1893, vol. iv., pp. 28 and following).

certain Australian tribes. Granting, he says, that circumcision, the removal of teeth, and other similar mutilations imply a condition of political or religious subjection—or both—no longer existing among these tribes, the custom is obviously the vestige of a more complex social condition. This conclusion seems the less reliable, since, according to the Rev. J. Matthew, the rite of circumcision was probably introduced into Australia by natives of Sumatra, and this view is confirmed by the local distribution of the custom and by other evidences of the same origin, such as the paintings which have been discovered in certain caves.[1]

(c) Teratology and social embryology also play a part in sociology, but it is of less importance than the others.

Certain customs among criminals show a resemblance to the habits of primitive man. On the other hand, we find cases where the individual development of an institution or society is a mere repetition of the development through which similar institutions and societies of other epochs and places have passed. Thus, for instance, there still exists in some parts of modern Russia a voluntary agricultural commune, for the periodical division of the land, an institution which existed more universally in the sixteenth and seventeenth centuries during

---

[1] *The Cave Paintings of Australia* (*Journal of the Anthropological Institute of Great Britain and Ireland*, April 1893, pp. 51 and following).

the existence and after the disappearance of family communities.[1]

In studying the development of these new communities, one can, in a measure, picture to oneself the development of laws of property, which were current in other countries at other times, and of which we possess little or no direct information. It is, however, hardly necessary to insist upon the hypothetical quality of such conclusions.

Our methods show, then, that organisms and societies exhibit considerable differences as well as analogies, a necessary result of their different natures. These few remarks must suffice; to add to them, we should have to overstep the limits we have set to this treatise, enter into well-worn controversies, and anticipate our own conclusions. Having merely explained our terminology, and indicated our general views, we will proceed to the subject of our investigations.

[1] *Tableau des origines et de l'évolution de la propriété et de la famille*, p. 170 (Kowalevsky).

BOOK I

UNIVERSALITY OF DEGENERATIVE EVOLUTION

PART I

DEGENERATION IN THE DEVELOPMENT OF INSTITUTIONS AND ORGANS

THE term "Evolution" does not in itself convey an idea of either progress or degeneration. It comprises all the changes undergone by an organism or society independently of the question as to whether these changes are favourable or otherwise. The evolution of an organ, or of the different parts of an organ, is degenerative if it tends to the ultimate decay of that organ or of its parts, and the facts are shewn by means of arranging series of fossils or living forms, and comparing them. Evolution is progressive if it tends to the development of an organ or to the formation of a new organ.

These definitions may be applied — *mutatis mutandis* — to the changes undergone by societies and institutions or their constituent parts.

The ideas of progress and of degeneration seem at first sight to relate to diametrically opposite phenomena. The term "progressive evolution" conveys the ideas of progress, development, improvement, of increasing differentiation, and of the progressive co-ordination of the functions or organs thus differentiated.

Degenerative evolution, on the other hand, conveys the ideas of decline, of decay, and of degeneration, such as the atrophy of the organs of locomotion in *Sacculina*, the degeneration of the leaves of parasitic plants, or the dissolution of corporate bodies in a declining state.

We propose to show, however, that these two ideas, which at first sight seem contradictory and mutually exclusive, are found, on a strict examination of the facts, to interpenetrate and complete each other. Degeneration and progress will appear as the two sides of one whole, or as two aspects of the same evolution, and it will be seen that all progress must necessarily be attended by degeneration.

CHAPTER I

IN THE EVOLUTION OF ORGANS ALL MODIFICATION IS NECESSARILY ATTENDED BY DEGENERATION

Section I.—Preliminary Considerations

In order to effect the demonstration which is the object of this chapter, only the phylogenetic modifications of organs will be discussed, setting aside the original formation of these organs in their ancestral forms, their development in the individual, the phylogenetic evolution of function, and the great variation resulting from individual adaptation.

Before entering upon our immediate subject, it would be as well to define its limits by saying a few words concerning the questions not under immediate consideration.

1. *Original formation of organs in ancestral forms.*—Little is known concerning the primary origin of organs, and their development before their assumption of the aspect and function by which we know them. What were the rudimentary leaves like in the ancestors of flowering plants? What were the eyes of the first vertebrates like? Did these organs develop from existing organs fulfilling other functions, or were they formed independently? However that may be, if they arose from other organs, we know nothing of the modifications which they underwent; and if they were formed independently, we need not discuss the fact here, for in that case the organ would not have developed by transformation. Having once been formed, it would develop and improve, and this process would not necessarily have been attended by partial degeneration.

2. *Development of organs in the individual.*—In the course of embryological development, organs do not exactly repeat the successive phases

through which they passed during their ancestral evolution. Generally speaking, and especially in the case of plants, the development of organs in an individual is direct, and gives no clue to its ancestral history. Moreover, when there is a recapitulation of ancestral stages, it often happens that evolution takes place without leaving traces of the various stages. This is especially the case in complex organs which have been produced by many lines of evolution converging in a single structure — a structure which thus becomes the seat of a special function or set of functions.

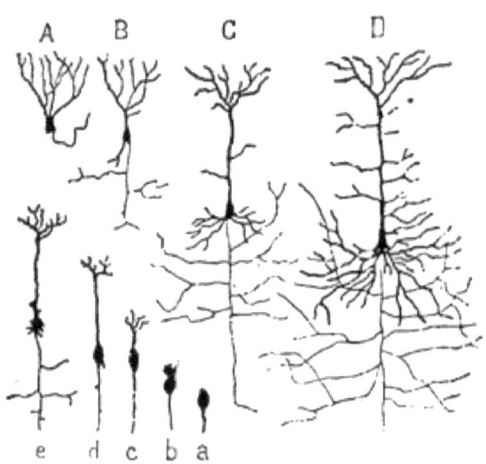

Fig. 1.—Diagram showing the evolution of pyramidal cells in the animal series.

The upper series of cells represents the psychic cell in various vertebrates: A, the frog; B, the lizard; C, the rat; D, man. The lower series shows the progressive stages in the evolution of the pyramidal cell in the human brain: a, the neuroblast without protoplasmic processes; b, the appearance of the nerve process and of the terminal ramifications; c, the nerve more fully developed; d, appearance of lateral branches of the axis cylinder; e, development of protoplasmic outgrowths of the protoplasm of nerve-cell and nerve. (Ramon y Cajal.)

The neuron, for instance, the ganglionic cell of the cortex of the human brain, passes successively through stages corresponding to those which are to be found in the adult fish, frog, bird, and mammal. In this case the development consists in an increasing complexity of the cell with no formation of unnecessary rudimentary parts.

3. *Phylogenetic evolution of function.*—Evolution may be regarded from a physiological, as well as from an anatomical standpoint,[1] but, in the former case, evolution is less a set of changes of function than an increasing specialization and division of labour, and under these circumstances it is often difficult to recognize a degenerative element in the evolution. A few examples will demonstrate this point:

Self-mutilation is a very common phenomenon among Echinoderms. Among brittle star-fish this reaction is controlled by some region of the nervous system; in some star-fish the reaction follows more quickly because the stimulus can act upon the ganglion at the root of each arm near the circumoral nerve-collar. In *Asteracanthum rubens*, there is a complete localization of this function, and self-mutilization only results when an exact region of the nervous system is stimulated.[2] In the Medusa we find an equally interesting example of functional evolution (Romanes).[3] With some of these (the

[1] *Evolution fonctionnelle du système nerveux.* J. Demoor, *Revue universitaire*, Bruxelles, 1892.

[2] *Contribution à la physiologie nerveuse des Echinodermes.* J. Demoor and M. Chapeaux. *Tijds. Ned. Dierk. Vereen.* (2) III. 2 Nov. 1891.

[3] *Preliminary Observations on the Locomotor System of Medusa; Jelly-fish, Star-fish*, and *Sea-urchins*. Romanes, *Int. Scient. Series*, 1885.

Acraspedote, on the outside of the umbrella being separated from the central part, the two separate parts continue to lash the water, the outer part with even strokes, the central mass more slowly and feebly. With the *Craspedote*, on the other hand, the central part, under the same conditions seems quite paralyzed and immovable, while the outer part continues to move in a perfectly normal manner. The causative function of the movement, the spontaneity of the movement as it was formerly called in physiology, is incompletely specialized in the *Acraspedote*, whereas in the *Craspedote* it is entirely localized.

Individual adaptation.—The individual is by no means a slave to heredity. It is capable of certain modifications under the influence of certain external conditions. These phenomena of individual adaptation may be arranged in three groups.

(*a*) When an organism, either animal or vegetable, is placed under new conditions of existence, when for instance, it relinquishes a terrestrial for an aquatic life, light for darkness, or fresh water for salt or estuarine water, its external aspect, and internal structure, undergo variations of considerable importance if it succeeds in adapting itself to the new conditions.[1]

[1] EXAMPLES: (*a*) The leaves of the water Ranunculus with laciniated leaves (*Ranunculus aquatilis fluitans*, etc.), are of normal structure when cultivated on dry land. The epidermis is furnished with stomata and the constituent cells contain no chlorophyll.

The organ does not, however, lose its primitive and typical characters. Actual organic transformation cannot, therefore, be said to take place in the case of individual adaptation.

The same leaves of the same plant when grown in water are much longer than those of the terrestrial type; the leaves have no stomata, and the epidermic cells are full of chlorophyll (Askenasy, *Ueber den Einfluss des Wachstumsmediums auf die Gestalt der Pflanzen*, Bot. Zeit., 1870, pp. 193 and following). Among the *Stratiotes aloides* it is not uncommon for the upper part of a leaf to rise above the surface of the water while the base is submerged; the epidermis at the base contains chlorophyll, but has no stomata, while the part of the same leaf which rises out of the water is furnished with stomata, but has no chlorophyll in the epidermis.

(*b*) A good example of this individual adaptation may be obtained by cultivating Cacti alternately in the light and in the dark. Goebel has shown that when a specimen of *Phyllocactus* is cultivated in the dark, the stems are prismatic and thorny; if the plant is afterwards placed in the light, the thorns disappear and the stems become quite smooth. (K. Goebel, *Ueber die Einwirkung des Lichtes auf die Gestaltung der Kacteen und anderer Pflanzen*, Flora, vol. 80, p. 96, 1895.)

(*c*) The animal kingdom furnishes numerous examples of individual adaptation.

The gradual drying up of Lake Aral caused the formation of a number of basins containing water at various stages of concentration. The Cardium of this region exhibits a whole series of adaptive variations. The shells become thinner and horny, their shape elongates, the openings contract, and their colour becomes duller (*Bateson*).

Mytilus edulis (the edible mussel), exhibits three different kinds of shells. It lives either in salt water, deep water, or shallow water visited by the tide. In each of these three vicinities the shells exhibit typical aspects.

The direction of boney lamellæ is known to agree with that in which the greatest strain is habitually applied, and the entire structure of a bone is dominated by the incidences of the forces applied to it. When, after a badly-mended fracture, the two broken (*b*) Our second group comprises the cases of cellular adaptation which may be produced in the course of embryonic life and may result in the formation of an embryo from half or the quarter of an egg.[1]

Here again the formation of organs takes place under new and abnormal conditions, but there is no "transformation" of these organs into fresh structures exhibiting other characters than those of primitive and typical organs.

pieces are joined by an oblique segment, the trabeculæ in this segment follow the direction from whence comes the greatest strain, a different direction to that which they would have taken had the inserted segment been placed parallel with the two broken pieces.

Without entering upon explanatory theories concerning these facts, it is important to notice that the adaptations of plants exhibit different characters to those of animals.

With plants it is only when a young organ is born under new conditions that it exhibits new characters. This is not necessarily the case with animals. In an animal organism the separation of the young tissues from the old is not so noticeable as with plants, as the organs undergo a continuous renewal and can always adapt themselves more or less to new conditions.

[1] When the two cells resulting from the first division of a fertilized egg of *Amphioxus* are artificially separated, each cell may develop directly into a complete individual. The same happens even when the first four cells are separated from each other artificially.

When this is effected in the case of *Echinus* or of an *Amphioxus* embryo of eight cells, each cell developes as if it had remained a part of the whole; but when the blastular stage is reached, that stage slowly completes itself.

In the first case the cell adapts itself to the new conditions, in the second case it is the blastular which does so. (See *Analytische Theorie der organischen Entwickelung*. Hans Driesch, Leipzig, 1894.)

In an adult cell, adaptation may, in a more or less normal way, cause the formation of new organs.

Even completely specialized cells of an organism may, when placed under certain conditions of life, adapt themselves to those new conditions and give rise to successive generations of different cells, thus generating other kinds of organs than those to which they would have given birth under normal conditions. This happens, for instance, with all the cells of *Salix*, which are capable of reproducing the whole organism by means of budding. This organic plasticity is shared with the *Begonia Rex*, but in that case it is only thoroughly developed in the epidermic cells. Analogous examples abound among animals, especially in the lower zoological groups.[1]

[1] Loeb has shown that in certain groups, when the organism, after having been wounded, is subjected to unusual light, position, or pressure, organs are formed at the wounded part, which are essentially different from those which would have been formed in a normal recovery. When this heteromorphosis occurs in *Tubularia mesembryanthemum*, *Aglaophenia pluma*, *Anthennularia rosa*, etc., the positions of the oral and apical poles may be transposed. When *Antennularia* is placed so that gravity acts upon it in a contrary direction to the ordinary one, there are formed oral and apical branches, where they would not have been formed had the organism been kept in a normal position. (Jacques Loeb, *Unters. zur physiol. Morphologie der Thiere*: I. *Ueber Heteromorphose*, Würzburg, 1891; II. *Organbildung und Wachsthum*, Würzburg, 1892.)

Similarly there is a true heteromorphosis when in a case of club-foot the faces and articular surfaces of the bones and cartilages assume characters adapted to the new work thrown on them as the abnormality of the joint increases. Another such case is in the false joints sometimes formed when the two

These various instances of cellular adaptation called " heteromorphosis " by Loeb, are outside the limits of our present researches, for reasons which we have already given.

The law of universal degenerative evolution applies only to the transformations of organs, and not to their original formation in either individuals or species, under normal or abnormal conditions.

Section I.

Transformation of organs of animals.

§ 2. *Transformation of homodynamic organs in the individual.*

In order to study the degenerative evolution exhibited in the specific modifications of the organs of an individual, it is necessary to choose homodynamic organs that are numerous, and consequently small enough to undergo different kinds of transformation fitted to the functions they may have to serve in the different parts of the body.

The numerous appendages of the cray-fish form an extremely interesting study of this kind. The

parts of a bone do not unite after fracture. In that case a completed joint with cartilage ligament and synovial membrane may be formed in the neighbouring tissues under the influence of the new stimuli. In extra uterine gestation the placenta is formed upon some abdominal organ. When this occurs, the unusual stimulus is sufficient to cause the cells of an abdominal organ to form a perfectly specialized organ of a nature foreign to their normal life.

appendages of the different metameres (the somites, or constituent segments of the body) of the crayfish are constructed on the same plan. The typical organ (fig. 2), when complete, consists of three parts—the *protopodite* (*pr.*), which is inserted into the body and carries a gill (*br.*); the *endopodite* (*en.*) and the *exopodite* (*ex.*), each of which consists of a series of joints which are attached to the end of the *protopodite* —the *endopodite* to the inner side, the *exopodite* to the outer.

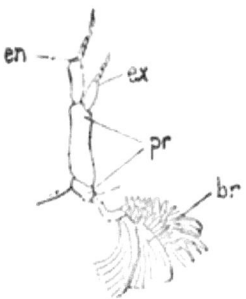

FIG. 2.—Diagram of typical and complete appendage of *Astacus fluviatilis*: *pr.* protopodite; *br*, gill; *ex*, exopodite; *en*, endopodite.

Let us now examine the twenty metameres of the animal in succession.

In each segment of the body we find a pair of appendages, the structure of which is based upon that of a complete appendage, but in which many different adaptations have brought about the conservation and increase of some of the typical parts, and the partial or total atrophy of other parts.

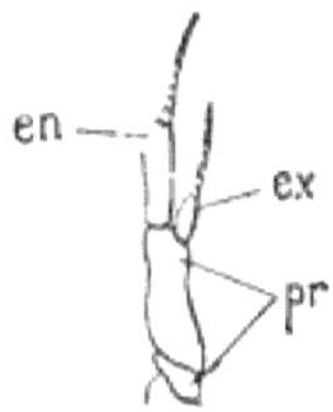

FIG. 3. — *Astacus fluviatilis*. Left appendage of the 3rd abdominal segment of the female (1,5/1): *pr*, protopodite; *ex*, exopodite; *en*, endopodite (Huxley).

Take first the abdominal segments, they carry appendages (fig. 3) formed of the three fundamental parts: the *protopodite* (*pr.*), the *exopodite* (*ex.*), and the *endopodite* (*en.*). Each of these parts is itself divided into different joints, the description of which is not of immediate importance. The portion of the appendage which has degenerated is the *podobranch*, which has completely disappeared. The reason is obvious; these appendages, having become a support for the eggs, have lost their respiratory function. The part specially adapted for that function has been allowed to atrophy, the more readily because the presence of gills on the ventral surface of the abdomen would be incompatible with free movement of that part of the body when the cray-fish is swimming.

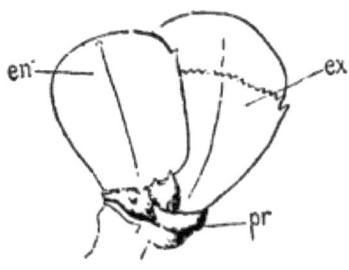

FIG. 4.—*Astacus fluviatilis.*—Left appendage of the 6th abdominal segment. Fins (1,5/1): *pr*, protopodite; *ex*, exopodite; *en*, endopodite (Huxley).

The appendages of the sixth abdominal segment are greatly modified into a caudal fin (fig. 4). This also exhibits the three fundamental parts of a complete appendage. The *protopodite* (*pr.*) is thick and short and has no gill, degeneration being exhibited in the loss of the gill, and by the reduced length of the part. The *exopodite* (*ex.*), and the *endopodite* (*en.*) are modified into two large oval plates which serve as propellers. The actual development of these two parts is accompanied by degeneration. In a typical appendage, the *exopodite* and the *endopodite* terminate in slender parts divided into several rings by false joints; in the propellers, which should offer the maximum resistance to the pressure of the water, the segmentation has disappeared, but across the part corresponding to the *exopodite* there still remains a transverse groove.

The second abdominal segment of the female carries appendages similar to those which have just been described. In the male, the organs of this segment, as also those of the segment in front of it, have become organs used in fertilization. We must consider what new structures have appeared here, and to what extent these new modifications have been attended by degeneration. The appendage of the second segment (fig. 5) is longer than that of the other segments; pressed against the ventral surface of the back part of the thorax, it stretches out as far as the space between the second and third walking legs, the part which corresponds to the oviduct of the female. It serves as a channel to conduct the fluid from the male orifice to that of the female. It consists of a *protopodite* (*pr.*) and an *exopodite* (*ex.*) similar to the corresponding parts of the appendages we have just described. The *endopodite* is profoundly modified. The inner border of its proximal region, which is not jointed, is extended into a thin plate rolled into a hollow horn (*a*) while the outer border is represented by an annulated part (*b*). The part in process of

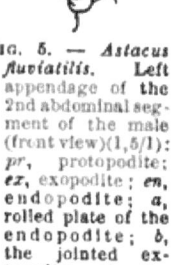

FIG. 5. — *Astacus fluviatilis*. Left appendage of the 2nd abdominal segment of the male (front view)(1,5/1): *pr*, protopodite; *ex*, exopodite; *en*, endopodite; *a*, rolled plate of the endopodite; *b*, the jointed extremity of the same (Huxley).

degeneration is the *exopodite*. On considering the development of the various parts of this appendage, it may be concluded that the *exopodite* has not undergone the same modification as the *endopodite*. It now remains to conclude this study of the abdominal appendages by an examination of the first segment. Immense variation occurs in the female. Sometimes there are two appendages, sometimes one or both are missing; in any case the existing organs are very small (fig. 6). The *protopodite* (*pr.*) is small, the *exopodite* is missing, and the *endopodite* (*en.*) is represented by an imperfectly jointed thread. The appendages of this first segment being no longer necessary, atrophy, and, as has just been pointed out, there is no regularity or uniformity in the order of their disappearance. In the male, the appendages of this segment (fig. 7) possess an unjointed rod (*t*) corresponding to the *endopodite* of the second segment, and this rod stretches out to a considerable distance. The *exopodite* is missing, and the articulations between the different parts have disappeared. This may reasonably be regarded as the result of degeneration, and confirms what has already been said concerning the appendage of the

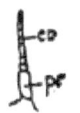

FIG. 6. — *Astacus fluviatilis.* Left appendage of the 4th abdominal segment of the female (3/1): *pr*, protopodite; *en*, endopodite (Huxley).

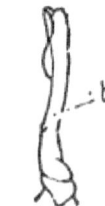

FIG. 7. — *Astacus fluviatilis.* Left appendage of the first abdominal segment of the male (1.5/1): *t*, unjointed rod (Huxley).

second metamere, where degeneration was exhibited by the relatively small size of the *exopodite*.

It may be concluded from this examination that in the abdominal appendages there is a degeneration common to all—this is the absence of the *podobranch*. In accordance with the particular adaptations of each pair of appendages, special degeneration accompanies special adaptive developments as we have shown above.

We now pass to the fourteen anterior metameres of the cray-fish. The first six of these constitute the head, and carry on the first segment the stalked eyes; on the second, the antennules; on the third, the antennæ; on the fourth, the mandibles; on the fifth and sixth, the two pairs of maxillæ. The eight metameres of the middle of the body form the thorax, and carry on the seventh, eighth, and ninth, the maxillipedes; on the tenth, the claws; and on the eleventh, twelfth, thirteenth, and fourteenth, the walking legs.

The structure of all these appendages may be referred to that of the typical primitive appendage. First take the thoracic appendages, commencing with the third maxillipede (fig. 8) which is the most complete appendage. It consists of a *protopodite* formed of two parts (*coxopodite* and *basipodite*) and carries a *podobranch* (*br.*), a well-developed *endopodite* (*en.*) consisting of five jointed parts, and of a small *exopodite* segmented like those of the abdominal appendages.

The third maxillipede may be regarded as the mean type of the average thoracic appendages. Compared with it the other appendages exhibit a true alternative development of their parts. When the *exopodite* is large the *endopodite* is small, and *vice versâ*. Besides these general modifications in which the degenerative evolution of the entire appendages is exhibited, each one of the segments undergoes certain modifications of its own.

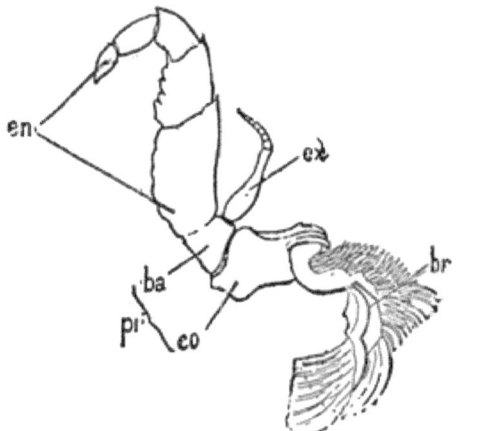
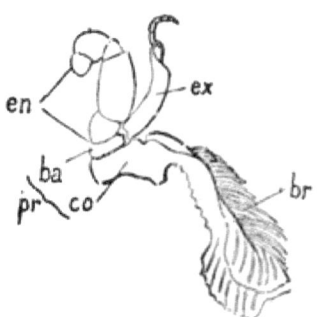

Fig. 8.—*Astacus fluviatilis.* Third left maxillipede (1,5/1): *co*, coxopodite, and *ba*, basipodite, forming *pr*, protopodite; *br*, gill; *ex*, exopodite; *en*, endopodite (Huxley).

Fig. 9.—*Astacus fluviatilis.* Second left maxillipede (1,5/1): *co*, coxopodite, and *ba*, basipodite, forming *pr*, protopodite; *en*, endopodite, *ex*, exopodite; *br*, gill (Huxley).

In the second maxillipede (fig. 9), which much resembles the third, the *exopodite* (*ex.*) is large, and the *endopodite* (*en.*) is small, the *prodopodite* (*pr.*) is better developed, and the *podobranch* (*br.*) has begun to atrophy.

The same evolution is exhibited by the first

maxillipede (fig. 10), but here it is more striking, and considerably modifies the general appearance of the appendage. The *exopodite* (*ex.*) is well developed, especially at its base; the *endopodite* (*en.*) is small and consists of only two joints; in the *protopodite* (*pr.*) the two component segments

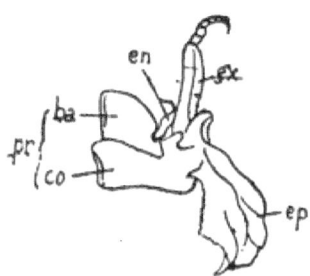

FIG. 10.—*Astacus fluriatilis*. First left maxillipede (1.5/1): *co*, coxopodite, and *ba*, basipodite, forming *pr*, protopodite; *en*, endopodite; *ex*, exopodite; *ep*, epipodite (Huxley).

are transformed into two long thin plates, and the *podobranch* is replaced by a membraneous plate (the *epipodite*) (*ep.*). Behind the maxillipedes are the claws and the four walking legs. In these five pairs of appendages the *exopodite* shows most signs of degeneration. We will now examine the component parts of these appendages (fig. 11).

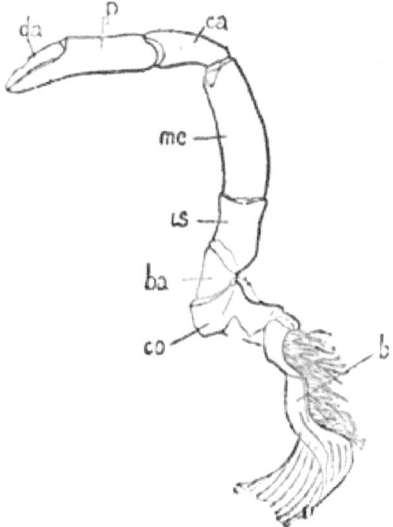

FIG. 11.—*Astacus fluriatilis*. Second left walking leg (1.5/1): *co*, coxopodite and *ba*, basipodite, forming *pr*, protopodite; *br*, gill; *is*, ischiopodite; *me*, meropodite; *ca*, carpopodite; *p*, propodite; *da*, dactylopodite (Huxley).

The *protopodite* (*pr.*) consists of two parts (*ba.* and *co.*), and carries a gill (*br.*) except in the last walking leg. The *endopodite* consists of five primitive joints, which are well developed and form the ordinary claw; these five parts are the *ischiopodite* (*is.*), the *mesopodite* (*me.*), the *carpopodite* (*ca.*), the *propodite* (*pr.*), and the *dactylopodite* (*da.*). The *exopodite* is missing. No vestige or rudiment of it is to be found in any phase of the development of the crayfish. In the lobster, however, which is closely allied to the cray-fish, the *exopodite* is still to be found during the larval period. The third fundamental part of the primitive member persists also in prawns throughout the entire period of life, but the organ is very small. At the extremity of the first and second pairs of walking legs there is an apparatus consisting of a fixed part—an elongation of the *protopodite*—and of a moveable part—the *dactylopodite*. This furnishes the walking leg with a prehensile organ which is well developed in the first pair of walking legs, and which is enormously increased in the true claws. In this evolution degeneration is exhibited by the disappearance of a joint, for in these appendages the *basipodite* and the *ischiopodite* are immovably united. This morphological degeneration corresponds to a functional change in the appendage. So long as the claw was used for locomotion a joint at this point was indispensable for progression. It is this joint which, in six or eight-footed beasts allows of the horizontal motion of the member necessary for locomotion, which in six-footed beasts results from the general

TRANSFORMATION OF ORGANS OF ANIMALS 39

structure of the parts and their auricular combinations. On the claw becoming prehensile, the joint hinge consolidates, the lever thus becoming much stronger and permitting the claw to be used to greater advantage.[1]

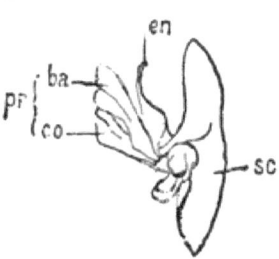

FIG. 12. — *Astacus fluriatilis*. Second left maxilla (1,5/1): *co*, coxopodite, and *ba*, basipodite, forming *pr*, protopodite; *en*, endopodite; *sc*, scaphognathide (Huxley).

It now remains to examine the appendages situated in front of the maxillipedes, *i.e.* the appendages of the head. In the second maxilla (fig. 12) a special transformation may be observed. The *coxopodite* (*co.*), and the *basipodite* (*ba.*), are flat plates; the *endopodite* (*en.*), which is small and undivided, exhibits signs of degeneration in its size, and in the absence of all articulations. The *exopodite*, according to some authorities on the subject, no longer exists, while according to others it constitutes with the *epipodite* (the analogous part to that which we regard as representing the gill in the maxillipede), a large peculiarly-shaped blade, the *scaphognathide* (*sc.*).

FIG. 13.—*Astacus fluriatilis*. First left maxilla (1,5/1): *co*, coxopodite, and *ba*, basipodite, forming *pr*, protopodite; *en*, endopodite (Huxley).

In the first maxilla (fig. 13), a partial degeneration of the organ is very marked: the *exopodite* and

---

[1] See J. Demoor, *Recherches sur la Marche des Crustacés* (Arch. de Zool. exp. et gen., 2° serie, t. iv., 1891).

T. List, *Bewegungsapparat des Arthropoden*, 1. Theil, *Astacus Fluviatilis*. Morphol. Jahrbuch., xxii. Bd. 3. Heft, 1895.

the *epipodite* are missing; the *endopodite* is reduced to a mere unjointed stem, and only the *protopodite* retains its two normal component parts.

The mandibles (fig. 14), the appendages of the fourth segment, are modified entirely for mastication. They consist of a strong transverse piece (*pr.*) provided at the extremity with an inner surface (*st.*) for grinding and sawing, and of a three-jointed piece (*en.*) with bristles which point outwards. The first piece is the result of the

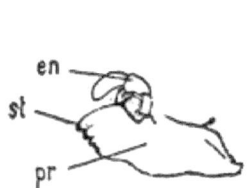 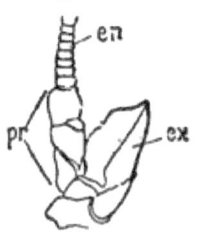 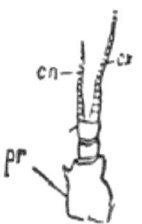

Fig. 14. — *Astacus fluviatilis*. Left mandible (1,5/1): *pr*, protopodite; *en*, endopodite; *st*, rasping surface of the protopodite (Huxley).

Fig. 15. — *Astacus fluviatilis*. Left antenna (1,5/1): *pr*, protopodite; *ex*, exopodite; *en*, endopodite (Huxley).

Fig. 16. — *Astacus fluviatilis*. Left antennule (1,5/1): *pr*, protopodite; *ex*, exopodite; *en*, endopodite (Huxley).

modification of the two parts of the *protopodite*, which have united to form the organ of mastication; the second, which represents the *endopodite*, is the feeler, an organ of sensation. All the other parts of the appendage have disappeared.

The antenna (fig. 15), which is a tactile apparatus, is formed of two parts representing the segments of the *protopodite* (*pr.*). The long-ringed process is the *endopodite*, while the lateral scale of the antenna represents a much reduced *exopodite* (*ex.*).

The antennule (fig. 16) consists of a *protopodite* (*pr.*), furnished with an annulated *endopodite* (*en.*) and *exopodite* (*ex.*).

The eye-stalk (fig. 17) consists of a two-jointed *protopodite* (*pr.*). This is all that remains, the *endopodite* and the *exopodite* being absent altogether.

This examination of the appendages of the cray-fish clearly shows that all fresh adaptation in the appendage entails the modification of some parts and the degeneration of others. In each case evolution is accompanied by degeneration.

FIG. 17.—*Astacus fluviatilis*. Left eye-stalk (1,5/1): *pr*, protopodite (Huxley).

§ 3. *Transformation of homologous organs in individuals of different species.*

The limitations of our present knowledge make it difficult to determine definitely the origin of limbs among vertebrates, but they are universally supposed to have developed from the lateral folds which still persist in *Amphioxus*, and which probably existed in the ancestors of vertebrates.[1]

[1] A. Morphological proofs:
    (*a*) Lateral folds of amphioxus.
    (*b*) Identity of the skeletons of paired and unpaired fins.
    (*c*) The number of spinal nerves passing to the fins.
    (*d*) The mode of entrance of these nerves into the fins.
B. *Embryological proofs:*
    (*a*) Continuous lateral folds in the embryoes of fish.
    (*b*) The formation of metameric pouches (coelomic invaginations not only at the point of origin of the limbs, but between them).

At first the only skeleton of these lateral rods consisted of parallel rods of a tough material. By the transformation of these rods, the skeleton of the limbs of vertebrates was ultimately formed.

The skeleton of limbs then consisted originally of a certain number of parallel rods. One of these rods, lying in the long axis of the future limb, became longer than the others, while the neighbouring rods began to slant, so that those nearest to the elongated rod spread out like a fan, and gradually moved outwards along the principal rod. This phenomenon was repeated several times, so that eventually those rods nearest to the principal rod passed towards the free end of it, and as the others followed in the same direction, the fin finally acquired a feather-like structure.

This transformation of the continuous folds into limbs—a progressive transformation, since new and more perfectly adapted organs were formed—was accompanied by degeneration, for a considerable part of the folds disappeared. In the same way, although the transformation of the parallel rods into bipinnated fins constituted a development,

C. *Palæontological proofs*:
    (*a*) The series of paired spines in *Diplacanthus* and in *Climatius* between the pectoral and ventral spines.
    (*b*) The skeleton of the fins of *Cladoselache*.

D. *Philogenetic proof*: The necessity of the bifurcation of the unpaired folds at the anal region (*Sternarchus*).

E. *Physiological proof*: The lateral folds are the undifferentiated condition of organs of equilibrium in modern fish.

the rods nearest to the free extremity of the principal rod were considerably reduced in size.

Two series of transformations should now be followed—that of limbs which have not yet ceased to be aquatic, and that of limbs adapted to terrestrial existence.

Among aquatic creatures, the Pleuracanthides and the Dipneusti (*Xenacanthus* and *Ceratodus*) have best preserved the bipinnated fin. In *Ceratodus* especially it is exhibited in almost the primitive condition. Gradually, however (as in *Orthacanthus*), the rods situated along one edge of the principal rod disappeared, and the fin, no longer bipinnated, became unilateral. This progressive transformation entailed the disappearance of nearly half of the fin.

With creatures which have become adapted to a terrestrial life, the limbs—so far as can be judged from what is known at present—appear to have undergone the following transformations: The bipinnated fin (such as that of *Ceratodus*) (fig. 18) is always the starting-point. Then the lateral rods of one half almost completely disappear (as in (*Protopterus Amphibius*). Next, the other half follows (as in *P. anncteus*) (fig. 19), and finally only the principal rod remains (as in *Lepidosiren*) (fig. 20).

At this point the limbs are reduced to mere lopped stems; they have not, however, atrophied; the degeneration which has accompanied this de-

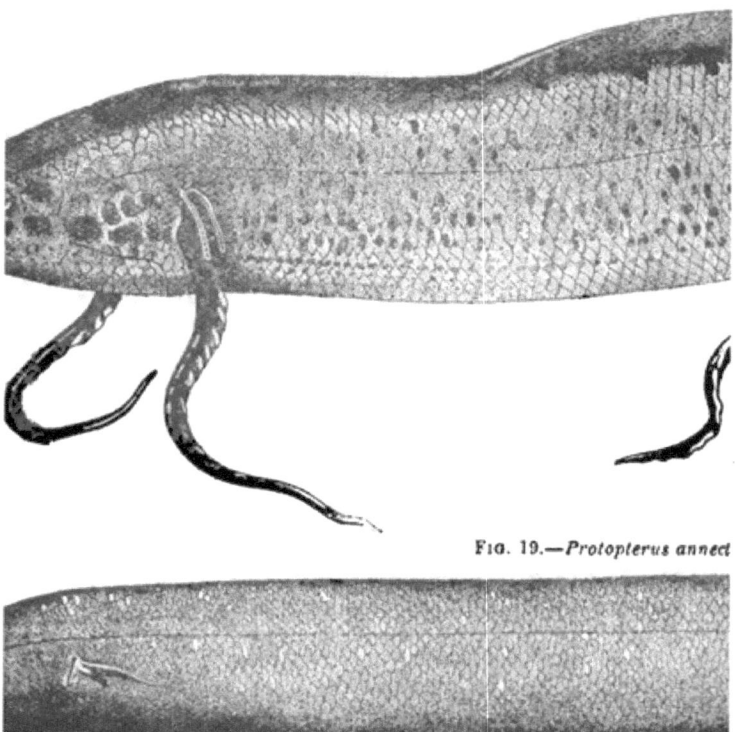

FIG. 18.—*Ceratodus Forsteri*, Krefft, Bipinnate fins (after Günther). (*See* DOLLO,

FIG. 19.—*Protopterus annect*

FIG. 20.—*Lepidosiren paradoxa*,

...logenie des Dipneustes, Bull. Soc., belge de Géol., de Paléont., et d'Hydrol., t. IX., 1895.

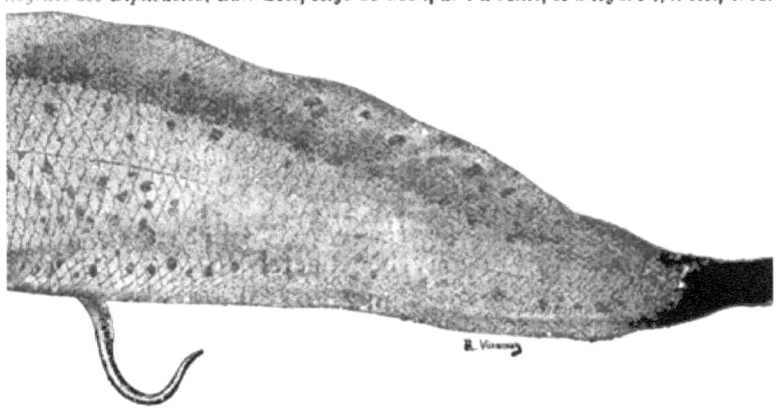

(after T. E. Gray). (*See* DOLLO.)

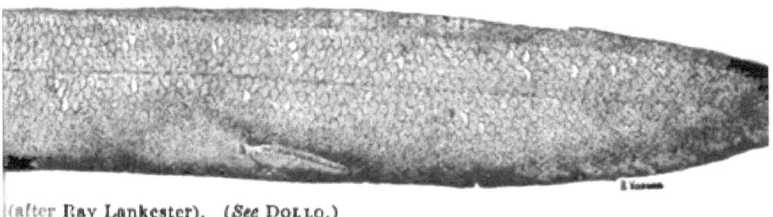

(after Ray Lankester). (*See* DOLLO.)

velopment has reached its ultimate limit; the limb has now to disappear altogether, or else to provide itself with new elements.

Judging from the embryology of the Urodele Amphibia, the primitive terrestrial limb had only one toe at first, and then a second appeared, then a third, a fourth, and a fifth. In this way was formed the primitive five-toed limb.

Let us now trace the adaptation of this five-toed limb to the principal functions it fulfils in the animal kingdom—walking upon two legs, running, jumping, climbing, burrowing, swimming, and flying. We shall find that the transformations it has undergone for all these functions have been accompanied by degeneration.

1. *Adaptation to walking on two legs.*—Adaptation to the upright position may be effected in two different ways, by walking or jumping. We will first investigate the former.

This adaptation to bipedal progression may be effected with or without the intervention of arboreal life, as may be seen by the following examination, the adaptations of man and of birds.

(a) *Man.*—The first sign, among mammals, of a definite adaptation to arboreal life, is the opposability of the great toe. The great toes of human beings are not opposable, the cases of "prehension" exhibited among savage races being due to lateral movements; we shall see presently, however, that this opposability of the great toe which is exhibited by all modern monkeys, probably existed at some time in man or his immediate ancestors.

The terrestrial five-toed foot must have acquired an opposable great toe in the course of adaptation to arboreal life. Without attempting to enumerate all the attendant phenomena of degeneration entailed by this modification, we will merely examine those exhibited by the nails of the hind foot. Comparison with the supposed ancestors of the Primates (Insectivores) shows that in the primitive terrestrial foot the nails completely covered the extremities of the terminal phalanges. With the lowest Lemur (*Tarsius*) the nails of the second and third toes are in the form of claws, while all the others are degenerating; they cover only the anterior face of the terminal phalanges, and are, in fact, flat nails. In all the other Lemurs, only the second toe has a claw, whereas in true monkeys all the nails are flat. Finally, in the Orang the great toe often has no nail at all.

We then come to limbs adapted to arboreal life, such as those of the Gorilla. With some anthropoids, however, and especially the Orang, the modification has gone further, and is accompanied by phenomena of degeneration in other parts of the limb. While, in the case of the Gorilla (fig. 21) the great toe has a true articulation between the metatarsal and the first phalanx, and also between the first and second phalanges, with the Orang (fig. 22) the great toe is much reduced so that it has no articulation between the metatarsals and phalanges, and frequently the last phalange is absent altogether.

The development of another foot, that of man, adds confirmation to the heading of this chapter. The great toes of man or his immediate ancestors must have been opposable. This may clearly be seen by a careful study of the muscles. We know what a resemblance there is between the muscles

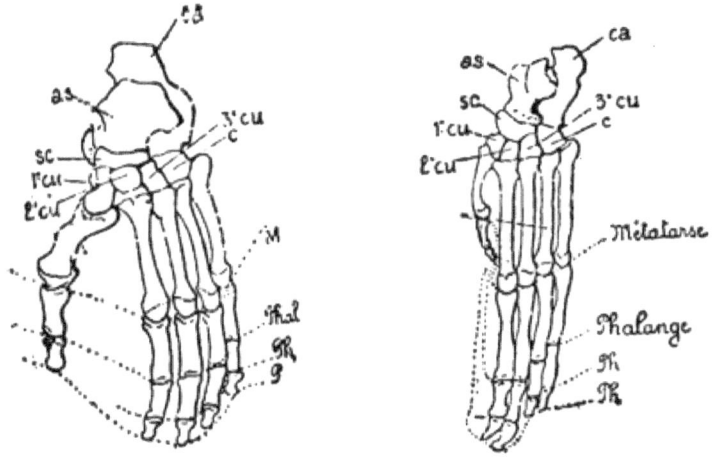

FIGS. 21 and 22.—Skeletons of the feet of the Gorilla and Orang-outang. *ca* calcaneum; *as*, astragulus; *sc*, scaphoid; 1re *cu*, first cuneiform; 2e *cu*, second cuneiform; 3e *cu*, third cuneiform; *c*, cuboid; *M*, metatarsals; *phal*, phalanges; *ph* and *p*, distal and terminal joints (after Waterhouse and Hawkins). See Huxley, *Man's Place in Nature*.

of the soles of the feet and those of the palmar region. With the exception of man, all mammals which have opposable thumbs, have also opposable great toes. On the other hand, there are various mammals (prehensile marsupials, rodents, bats, marmosettes, etc.) which have opposable great toes but not opposable thumbs. We may presume then that the hind limbs were the first to become adapted to arboreal life, and, since the great toe of man is not opposable, it may be concluded that it ceased to be so when terrestrial life was assumed. This modification is not wholly degenerative, for the functions of the lower limbs are no less important in man than in apes. There are, however, undoubted signs of degeneration in the changes entailed. The muscles which rendered the great toe opposable were probably the first to degenerate, but by referring to a skeleton it may be seen that whereas the great toe and the metatarsals have gained in length, the first phalanges, and still more the other phalanges, are reduced. In some cases the last phalange of the little toe has even disappeared altogether.

(b) *Birds.*—First take the reptile-footed Dinosaurians. These are four-footed beasts, and the hind limbs are plantigrade and five-toed. Counting from the inner side the phalanges of the toes are: 2, 3, 4, 5, 4 in number.

In passing on to the bird-footed Dinosaurians we find that with *Hypsilophodon*, the foot is functionally four-toed. From the first to the fourth toe the phalanges are the same in number as with the reptile-footed Dinosaurians, but the fifth toe is greatly reduced, being merely represented by a pointed metatarsal. In *Camptosaurus*, the fifth toe has disappeared, the second, third, and fourth are fully functional, but the first, although retaining the usual number of phalanges, is greatly reduced in size. Finally, in Iguanodon,[1] which is fully adapted to the upright position, the second, third, and fourth toes are still fully functional, but not only has the fifth toe disappeared, but the first is merely represented by a pointed metatarsal.

Thus it is apparent that each new phase in the transformation from quadruped to biped is attended by partial degeneration, although the hind limbs, having alone to fulfil the function of locomotion, necessarily acquire a fuller development. The *Iguanodon* is too specialized in its development to have been the ancestor of birds; but this is not so with other bird-footed Dinosaurians, such as *Camptosaurus* and *Hypsilophodon*, where the hind foot is sufficiently primitive to have been the forerunner of the bird's claw.

Most birds have four functional toes, which have retained the primitive number of phalanges, but the legs of running birds exhibit more or less important variations. The modifications they have undergone have obviously not diminished their functional importance, but quite the contrary. These modifications have, however, been accompanied by degeneration with running birds, and the first toe disappeared first, as in the *Cassowary* and the *American Ostrich*,

[1] Dollo: First, second, third, fourth, and fifth *Notes sur les Dinosauriens de Bernissart*. *Bull. Mus. royal d'Hist. nat. de Belg.* 1882, 1883, 1884.

to be followed in the second stage by the second toe. Then the fourth began to atrophy until the whole weight of the body was supported by the third toe alone, as in the ostrich.

2. *Adaptation to leaping*—We now come to the various instances of adaptation to leaping. This adaptation is one necessarily undergone either by an animal which originally lived a terrestrial life, such as the Jerboa, or by one which had ceased to be arboreal in adapting itself to leaping, such as the kangaroo; or, again, by an animal which formerly lived an arboreal life, and which continued to lead such a life with the additional adaptation to leaping, such as the tarsius; and, finally, by an animal which first lived and swam in the water, and which continued to do so after becoming adapted to leaping, such as the frog.

In each of these cases adaptation to leaping has resulted in the fuller development of the hind limbs, the size of which, as well as the functional importance, has greatly increased, which shows that the modification is of a progressive nature. All these modifications are, however, attended by degeneration.

(*a*) *The Jerboa* (*Dipus*).—Some rodents, such as rats, porcupines and squirrels, are five-toed. This five-toed foot represents the primitive one from which, after passing through various still existing phases, was derived the three-toed foot of the jerboa.

With the hare there is no first toe, so that the foot is functionally four-toed.

The fifth toe next disappeared, leaving the foot three-toed, as in the viscacha, the capybara and agouti. In the foot of the agouti the metatarsals have considerably gained in length, and are closely pressed together.

In the jerboa (fig. 23) the three metatarsals (*M.*) are united, and the intervening tissues have disappeared. Besides this modification, the hind limbs of the animal, being used for leaping only, have become thinner, while gaining in length by the elongation of the metatarsals.[1]

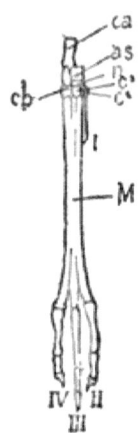

Fig. 23.—Hind limb of *Dipus* (Jerboa). I, first rudimentary digit; II, second digit; III, third digit; IV, fourth digit; *M*, bone formed by fusion of 2nd, 3rd, and 4th metatarsals; *ca*, calcaneum; *as*, astragulus; *n*, navicular; *c³*, third cuneiform; *c²*, second cuneiform; *cb*, cuboid. (After Flower, *Introduction to the Osteology of the Mammalia*.)

(*b*) *The Kangaroo.*—Although now inhabitants of the plain, kangaroos are descended from arboreal marsupials, which themselves have gradually ceased to be arboreal.[2] All the intervening phases of this transition from type to type are known.

[1] The last segment of the leg of the jerboa—the foot being formed by the union of the second, third and fourth metatarsals—corresponds more or less to the formation of the bird's foot. This, however, is a case of convergent modification more apparent than real, for in the cannon bone of birds, besides the second, third and fourth metatarsals, there is the whole set of distal tarsals, whereas in the jerboa the distal tarsals remain free and isolated bones.

[2] L. Dollo, autographs of the course given at the Institute of Solvay, 1891-1892, 5th lecture, *The Origin of Kangaroos*.

In the Opossum (*Didelphys*) we find a normal five-toed foot, in which the third toe predominates, but which is fully adapted to arboreal life, the first toe being opposable and provided with a flat nail.

The foot of the Phalanger (*Phalangista*) (fig. 24) is also five-toed, and the great toe is opposable, but here it is the fourth toe (iv.) which predominates owing to the degeneration of the second (ii.) and third (iii.)

The Kaola (*Phascolarctos*) exhibits the same type of foot, but with an even more marked diminution of the second and third toes.

In *Hypsiprymnodon* the first toe has become rudimentary.

In *Perameles* the first toe is only represented by a metatarsal with phalanges.

In the kangaroo (fig. 25) it has completely disappeared.

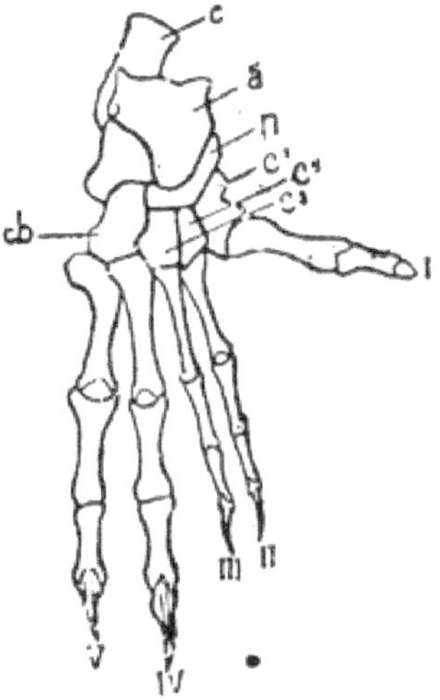

FIG. 24.—Right fore-paw of Phalanger.
c, calcaneum; *a*, astragulus; *n*, os centrale; *cb*, cuboid; c^1, c^2, c^3, first, second, and third cuneiforms; I, II, III, IV, V,—1st, 2nd, 3rd, 4th and 5th digits, of which the second and third are degenerating. (After Flower, *Introduction to the Osteology of the Mammalia*.)

Finally, in *Chæropus* (fig. 26) the fifth toe is reduced to a mere thread, as are also the second (ii.) and third (iii.) toes.

This last stage, which represents the present and most perfect adaptation of marsupials to leaping, was attained by the predominance of the fourth toe, accompanied by the almost total extinction of the other toes. The persistence of the fourth toe

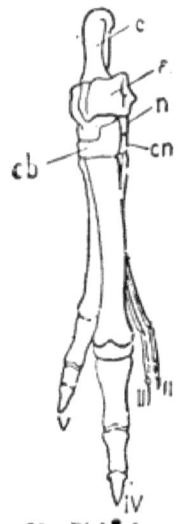
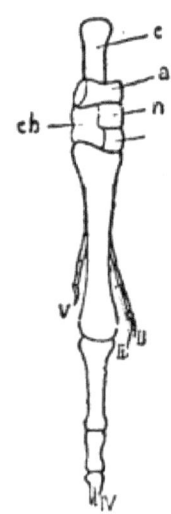
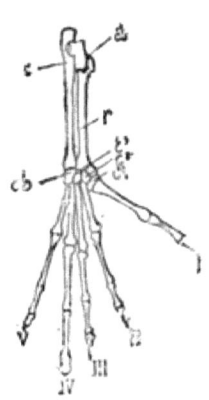

Fig. 25.—Right fore-paw of Kangaroo. *c*, calcaneum; *a*, astragulus; *cb*, cuboid; *n*, os centrale; *cn*, third cuneiform. II, III, IV, V,—second, third, fourth and fifth digits, the second and third being degenerate. (After Flower's Introduction to the *Osteology of the Mammalia*.)

Fig. 26.—*Chœropus*. Right fore-paw. *c*, calcaneum; *a*, astragulus; *n*, os centrale; *cb*, cuboid; c^3, $3e$, third cuneiform; II, III, IV, V,—second, third, fourth and fifth digits, of which the second, third and fifth are degenerating. (After Flower's Introduction to the *Osteology of the Mammalia*.)

Fig. 27.—Right hind-paw of *Tarsius Spectrum*. *c*, calcaneum; *a*, astragulus; *r*, navicular; c^1, c^2, c^3, first, second, and third cuneiforms; *cb*, cuboid; I, II, III, IV, V,—first, second, third and fourth digits. (After Flower's Introduction to the *Osteology of the Mammalia*.)

instead of the third is accounted for by the previous arboreal existence of marsupials before they became adapted to leaping.

(*c*) *The Tarsius* (*Tarsius spectrum*).—The tarsius, which is a small lemur, is to be found in Celebes.

It is so-called owing to the remarkable length of its tarsus (fig. 27), a very exceptional formation, for the elongation of an animal's foot is usually effected by means of a lengthened metatarsal. This abnormal formation in the foot of the tarsius can, however, be accounted for. Like most of the Primates, the tarsius is adapted to an arboreal life, and, among other characters, it exhibits an opposable great toe. When on the ground, however, instead of running on two or four legs, it progresses by leaping; and this may also be noticed with some of the lemurs of Senegal (Galago). The tarsius has become adapted to leaping by the lengthening of the hind limbs, the thigh, leg and foot being equally elongated. The foot, however, being of fan-like structure, it is clear that if all the toes were elongated, the feet would overlap one another in the middle line, and therefore with most animals that leap the elongation of the foot involves a reduction in size of the outer and inner toes to prevent overlapping. The tarsius, however, has remained arboreal, although it has taken to leaping, and it therefore requires a well-developed foot with an opposable great toe in order to obtain a firm grip on branches. To avoid overlapping of the feet, the length has been augmented at the ankle, which, not being shaped like a fan, does not consequently gain in length.

A more detailed examination would doubtless

show that this modification of the foot has been attended by the degeneration of some parts of it, but, as we have to restrict ourselves here to the data with which we are already furnished, it is the degenerative phenomena exhibited by other parts of the limb consequent to adaptation to leaping, which we shall proceed to enumerate.

There are two leg bones, the fibula, and a larger one, the tibia. These two bones are usually held together by muscles, or by an interosseous membrane, the latter being merely degenerated muscular tissue. As a rule, the tibia and the fibula reach the entire length of the leg and are quite separate. In the process of adaptation to arboreal life, these bones have remained apart, as may be observed in most arboreal animals of the present day. Moreover, in order to promote prehension by the foot, the tibia and fibula in prehensile-footed marsupials may turn upon one another, as is the case with our radius and ulna. With the tarsius it is quite otherwise. Here the lower halves of the tibia and fibula are united, and the muscles which formerly connected the two bones have consequently disappeared. There is further evidence of degeneration in the upper half of the fibula which has become reduced to a mere thread. There is no question that these phenomena are the result of adaptation to leaping, for they are not exhibited among exclusively arboreal animals, while many parallel cases of modification are known to exist amongst other species.

(d) *The Frog* (*Rana Esculenta*).—Adaptation to leaping entails, as a rule, the elongation of the hind limbs. When a large blade-like swimming foot is required, the number of toes cannot be diminished and length must be augmented, as in the tarsius, in the region of the ankle (fig. 28). But, in the frog, it is the astragulus (*a.*) and the calcaneum (*c.*) which have elongated, whereas in the tarsius it was the calcaneum and navicular bones.

Besides the union of the calcaneum and the astragulus bones—which may be regarded as a phenomenon of degeneration, as it entails the disappearance of the intervening tissues—the adaptation of the frog to leaping has been attended by a yet more characteristic sign of degeneration: the tarsus, which is a very complicated structure in tailed amphibia, is reduced, especially in the distal row, to a few minute bones.

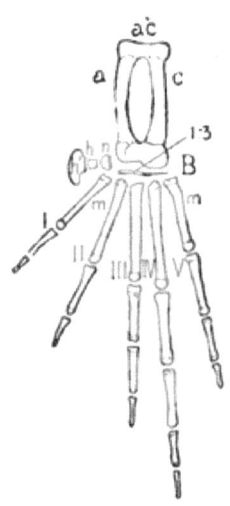

FIG. 28.—Right foot of *Rana esculenta.*
a, astragulus; *c*, calcaneum; 1-3, *n*, *h*, *h*¹, tarsals; *m*, metatarsals; I, II, III, IV, V, digits. (After Ecker, *Anatomy of the Frog.*)

3. *Adaptation to running.*—The ruminants and the horse furnish us with two excellent examples among mammals of adaptation to running.

Their common ancestor is known to have been a

clumsy five-toed animal which lived in swamps. As the horse type was evolved, the legs, which originally only served to support the weight of the body in a slow progress, were gradually adapted to running. This modification was certainly advantageous to the animal, as it enabled it to escape more easily from its enemies. It was, however, accompanied by degeneration, for the ancestor of the horse lost first the great toe, then the fifth, and next the second and third toes became so reduced that eventually only one toe remained functional.

Adaptation to running was effected differently with the ruminants, but it was equally accompanied by degeneration. With them, first the great toe disappeared, then the second and fifth became simultaneously much smaller, while the third and fourth continued to increase equally in length. In process of time the second and third toes entirely disappeared, and the metatarsals of the third and fourth joined together to form the cannon bone. In this way mammals were evolved with cloven hoofs and adapted to running.

4. *Adaptation to flying.*—There are three types of animals among the vertebrates: birds, pterosaurians, and flying bats, by which is meant, not those bats which are merely able to float in the air like a parachute, but those which can both raise and guide themselves during flight.

(*a*) *Birds.*—We have seen that the ancestor of all terrestrial vertebrates was a five-toed animal.

The hand of the oldest known bird, however,—*Archæopteryx*— had only three digits, and the question arises as to whether this atrophy of the two outer digits was effected in the process of the animal's gradual adaptation to flight or if it dates from a still earlier period.

Archæopteryx had four toes. The fore legs were smaller than the hind legs. Now, with bipedal or quadrupedal animals which make no special use of their fore legs, the contrary may invariably be noticed. It is of small importance, therefore, whether *Archæopteryx* was quadrupedal or bipedal before it took to flying, for neither of these modes of progression would have entailed the loss of two of its digits, and the disappearance may reasonably be attributed as due to the process of adaptation to flight.

The three remaining digits of *Archæopteryx* were complete in their structure. The phalanges, which were of different lengths, are furnished with claws, and were of the same number as is normal among reptiles, *i.e.* 2, 3, 4, counted from the inner side. The wing of *Archæopteryx* had not therefore undergone much modification beyond the toes of the two outer digits, but it must be remembered that the bird was a weak flyer; instead of having a large boney sternum necessary for powerful flight, the sternum of *Archæopteryx* was formed of cartilage, and the wings were short and rounded in shape.

With good flyers, on the other hand—such as pigeons or sea-birds—the adaptation to flight is more perfect, the modifications entailed are much greater and have been accompanied by further signs of degeneration. In the fowl (fig. 29) all the claws have disappeared (*degeneration No.* 1); the thumb (i.) and the index finger (ii.) have each lost a phalanx; the middle finger (iii.) has lost three (*degeneration No.* 2); finally, the metacarpals of the index and middle fingers have fused (*a.*) thus entailing the degeneration of the muscles between those parts (*degeneration No.* 3).

(*b*) *The Pterosaurians.*—These reptiles, which are now extinct, were able to fly, like bats, by means of a membrane. There is no connection between them and birds which fly in quite a different way.

Fig. 29.—Skeleton of wing of fowl.
I, First digit or thumb; II, second or index digit; III, third or great digit; *a*, fused metacarpals of the second and third digits. (After Huxley.)

The membrane of the Pterosaurians (fig. 30) was supported by a framework of which the most important part was the fifth finger (v.), the digit corresponding to our little finger, which was greatly elongated.

The patagium (the flight membrane) consisted of two kinds of membrane, the antebrachial membrane which extended from the neck to the wrist, and the true wing membrane, which reached from the free end of the little finger to the ankle.

There are three essential types of Pterosaurians. *Pterodactylus*, which had a short tail and teeth all along the gape; *Rhamphory-hnchus*, which had a long tail and a beak set in the front part of the jaw with teeth behind it; and *Pteranodon*, which had neither teeth nor tail. This last type was the most specialized of the Pterosaurians,

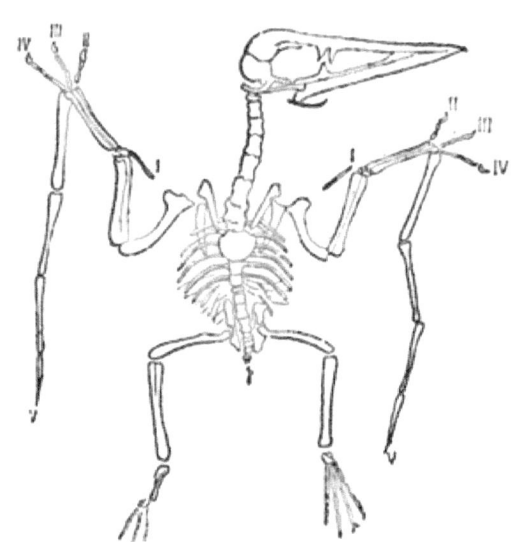

FIG. 30.—*Pterodactylus spectabilis* (after Huxley, *Anatomy of Vertebrated Animals*).

and the strongest flyer of the whole group. They attained to a huge size, the skull measurement averaging 1 metre 20, and the wing extension 8 metres 30. The modification which the fore-limbs of the Pterosaurians underwent in the course of their adaptation to flight, was accompanied by degenerative phenomena in proportion to the extent of the specialization (see *Pterodactylus spectabilis*) (fig. 30).

In the process of adaptation to flight the thumb (i.), and the little finger (v.) first lost their nails and became useless. Next, the metacarpal and the proximal phalange of the thumb became much smaller, and united with the bones of the wrist, while the distal phalange degenerated to a mere thread. Fresh evidences of degeneration presented themselves as the specialization became more advanced. In *Pteranodon*, for instance, the metacarpals of the clawed digits were considerably reduced in size, the arms having almost ceased to exercise any function but that of flight. There was a special provision for the support of these great wings,

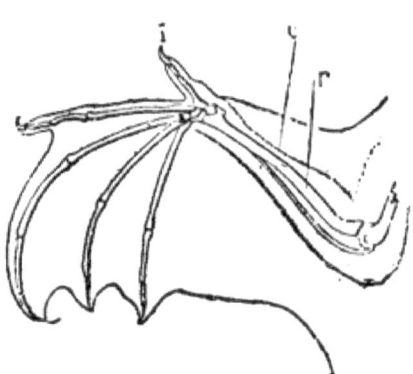

FIG. 31.—*Pteropus* (a bat). Skeleton of fore-limb; *r*, radius; *c*, degenerating ulna; I, II, III, IV, V, digits. (After Huxley, *Anatomy of Vertebrated Animals*.)

the development of which was only equalled among vertebrates by that of the pectoral fins of the skate. Whereas the legs were upheld by the pelvis, which was in its turn supported by the sacrum, the arms of *Pteranodon* were attached to the middle of the shoulder blade, the latter being supported by the vertebral column—an absolutely unique condition. This kind of pectoral sacrum implied the fusion of a number of vertebrae, and involved a consequent degeneration of the muscles which in earlier Pterosaurians caused the movement of one vertebra upon another.

Two phases may thus be observed in the evolution of the Pterosaurians: first, *Pterodactylus* and *Rhamphorynchus*, and next *Pteranodon*, while degeneration attended the modifications of both.

(c) *Bats.*—Bats, like the Pterosaurians, fly by means of a membrane, only that the membrane, instead of principally extending from the little finger to the body, is equally developed between all the fingers.

Let us consider what degeneration is involved by this modification.

Firstly, all the digits have lost their nails, excepting the thumb and the index finger in fruit-eating bats, and the thumb alone in insect-eating bats, while some of the phalanges of the digits are missing, the usual number being two instead of three. Further, instead of the ulna and radius being equally developed in the fore-arm, the ulna is greatly reduced in size.

5. *Adaptation to arboreal life.*—We have seen that, as a rule, the first evidence of adaptation to arboreal life is the opposability of the great toe. This modification is accompanied by the degeneration of the nail, for, instead of consisting of a claw covering the entire extremity of the last phalange, it becomes a small nail covering only the upper side.

Another modification is exhibited at a slightly

later stage. In the primitive foot of the aminota, the third toe is the longest, but among animals which have become adapted to an arboreal life (fig. 32) the fourth toe preponderates, thus allowing of a wider grasp, and the second (ii.) and the second and third toes gradually degenerate (*Arctocebus calabarensis*, Potto). A clear proof that this modification is due to adaptation to arboreal life, lies in this same tendency being exhibited by two groups not closely allied to one another—the marsupials and the lemurs.

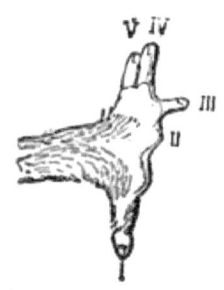

FIG. 32.—*Arctocebus calabarensis*. Right hand. I, II, III, IV, V = the five fingers, of which II and III are degenerating. (After Huxley, *Anatomy of Vertebrated Animals*.)

In a still more advanced stage the phalanges of the second digit disappear almost completely, and, for all practical purposes, no longer exist.

Among reptiles, chameleons furnish a striking example of adaptation to climbing. With them three digits are opposable to the other two (figs. 33 and 34), instead of one being opposable to the other four. To further promote this opposability, there is a fusion of digits (syndactilism), *i.e.* each set of digits is enclosed up to the nails within a common integument. The result of this is that the enclosed digits are no longer capable of lateral movement, and the muscles have degenerated in consequence. The wrist and tarsus also show signs of partial degeneration.

6. *Adaptation to swimming.*—Among mammals, adaptation to swimming entails the functional disappearance of the lower limbs, and the modification of the fore-limbs into fins.

This modification involves considerable degeneration. The alteration in the number of digits varies; the whale-bone whales have kept all five; in rorquals only four remain; and, after a series

 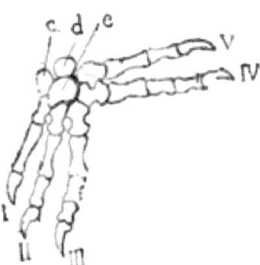

Fig. 33.—Skeleton of hand of Chameleon. (After Cuvier.)

Fig. 34.—Chameleon. Skeleton of foot. (After Cuvier.)

of intermediate conditions, the dolphin is evolved, the fins of which are functionally two-toed. If the Cetaceans vary as to the number of digits they exhibit, they all have this character in common—the joints have disappeared. The arm no longer articulates at the elbow, and neither the wrist nor the phalanges are jointed. The simple stratum of cartilage which takes the place of the true joints allows of a slight movement of the various segments, but not of true articular movement.

This results in the degeneration of all the muscles which caused this mobility in their terrestrial ancestors.

A similar degeneration may be observed among Plesiosaurians, Ichthyosaurians, and Mosasaurians.[1]

With regard to Sirenia, the Dugong is unable to move the arm at the elbow joint, but the Manatee can grasp things with its fins. Besides many other modifications, the adaptation of these animals to an aquatic life has entailed the fusion of the two bones of the forearm, thus involving the degeneration of those muscles which formerly prevailed over the movement of the two bones when separate.

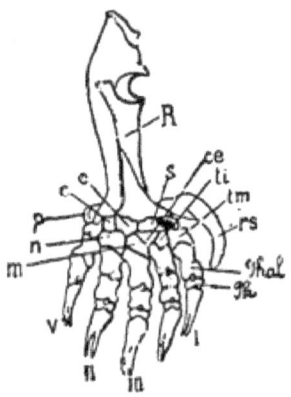

Fig. 35.—Fore-limb of *Talpa europaea* (the Mole).
R, radius; s, scaphoid; c, semilunar; c, cuneiform; p, pisiform; n, unciform; m, os magnum; ti, trapezoid; tm, trapezium; ce, os centrale; rs, falciform sesamoid; phal. phalanges; ph, distal joints of digits. (After Flower.)

7. *Adaptation to burrowing.*—There are two types of adaptation to burrowing —that of the Mole and that of the *Heterocephalus*.

In the mole (fig. 35), the fore-limb, although it has become shorter in order that it may better fulfil

---

[1] L. Dollo, *Première note sur les Mosasauriens de Mesvin* (*Bull. Soc. belg. Géol. Paléont. Hydrol.*, vol. iii., 1889); *Nouvelle note sur l'ostéologie de Mosasauriens* (*Bull. Soc. belg. Géol. Paléont. Hydrol.*, vol. v., 1892).

its function, is obviously in process of development, for it has not only retained all five digits, but, what is functionally a sixth digit, has made its appearance (rs.). The phalanges are very well developed (phal.), but, on the other hand, the distal phalanges (ph.) have degenerated and are very short.

Heterocephalus (fig. 36) is a burrowing rodent, only, instead of being talpoide as is the *Bathyergus*, *i.e.* instead of exhibiting a body furnished with

FIG. 36.—*Heterocephalus Philippi*. (After Oldfield Thomas.)

strong short legs, *Heterocephalus* looks more like an ordinary quadruped. In process of adaptation to an underground life, the hair, especially that on the legs, has disappeared. This is the more remarkable, as in no other case does it occur among mammals unless as an adaptation to aquatic life. Here it is in the atrophy of the roots of the hair that degeneration is manifested.

One form of hair, however, is still to be found on *Heterocephalus*, for there are a few bristles on the outer sides of the feet. These serve as brushes to sweep away the sand while burrowing.

Section II.

Modification of the organs of plants.

§ 4. *Modification of homodynamic organs in the Individual.*

For the same reasons as we have already pointed in the case of vertebrates, those leaves which grow out of different parts of the main stem, and have undergone many modifications in order to adapt themselves to their various functions, will best serve as examples for our present demonstration.

In descriptive botanical treatises the term "leaf" is applied without discrimination. As a rule, however, this term should only be applied to the true leaves, *i.e.* those of which the parts are fully developed throughout.[1]

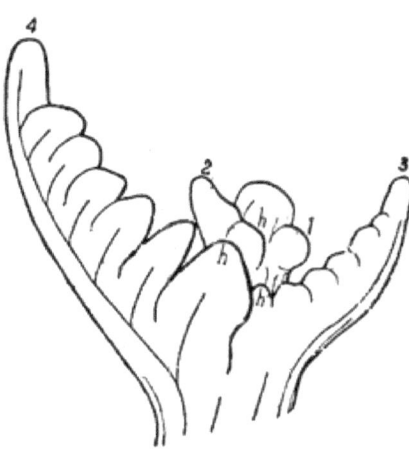

FIG. 37.—Growing point of *Vicia varia*, showing the origin of leaves.
1, 2, 3, 4, leaves respectively older; leaf 1 is still in the primordial condition; leaf 2 is divided into epipodium and hypopodium (*h*). In leaves 3 and 4 the epipod is divided into petiole and leaflets, and the hypopod forms stipules.

[1] The embryological development of the leaf among the Angiosperms (fig. 37) exhibits the following phases :—The leaf arises from the growing point in the form of a small bud, which grows

These foliage leaves, which are solely assimilative, are philogenetically the oldest, for the assimilative function of leaves was certainly an earlier function than any they may now exercise; but it is not uncommon to find both basilar (*Niederblätter*) and apical leaves (*Hochblätter*) as well as foliage leaves (*Laubblätter*) in the same plant, and even growing from the same branch. These leaves, which arise directly from the foliage leaves, have more or less completely lost their primitive function, and have assumed others. Their structure has undergone corresponding modifications which are all attended

for some time without any evidences of differentiation, and constitutes what is called the primordial leaf. Next the leaf is differentiated into a proximal part, which almost surrounds the stem (the hypopod), a distal part (the epipod), and an intermediate part (the mesopod). From this point the various parts of the leaf begin to develop individually.

Little foliacious lamina (the stipules) frequently arise laterally from the hypopod, and, while still in the bud these stipules, being often much larger than the rest of the leaf, serve as a very effective protection to the young organs.

The way in which the epipod differentiates varies very much in different cases. The lamina is developed from it, and little buds are gradually formed around the point until the whole is ready to branch out. At this stage the young leaf is still rolled up within itself, and protected by stipules, when there are any.

Next, the mesopod grows into the petiole, and the growth of the petiole causes the separation of the different parts of the leaf.

The leaves of many plants are much less complicated than these. Sometimes the mesopod is missing, and the hypopod and the epipod are left in contact. Sometimes there is practically no hypopod, and in some cases the primordial leaf develops without any differentiation into hypopod and epipod. See fig. 49 (*Sempervivum arachnoideum*).

70 UNIVERSALITY OF DEGENERATIVE EVOLUTION

by degeneration. A good demonstration of this will be to examine the leaves of the following types: *Rosa rugosa*, *Serratula centauroides*, *Sagittaria sagittifolia*, *Lathyrus Aphaca*, and *Nymphaea dentata*.[1]

1. *Rosa rugosa*.—In the foliage leaves (fig. 38, F) of a rose-branch, such as of *Rosa rugosa*, there are two

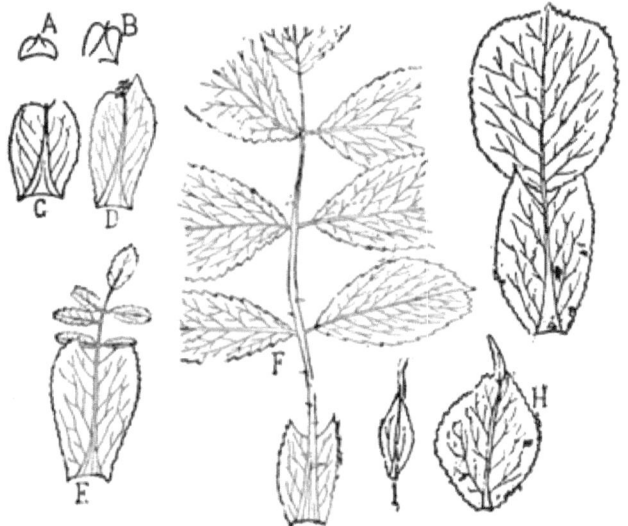

Fig. 38.—Leaves from the same branch of *Rosa rugosa*.
A—E, successive basilar leaves; F, foliage leaf; G—I, successive apical leaves.

lateral stipules which arise from the hypopod and are fused with the base of the petiole. The petiole, which arises from the mesopod, has from two to six pairs of leaflets, and also a terminal leaflet—all of which are derived from the epipod.

[1] In studying these types we will confine our attention to the leaves arising directly from the foliage leaves, to the exclusion of the floral leaves.

MODIFICATION OF THE ORGANS OF PLANTS 71

The basilar leaves (fig. 38, A to E), which are usually withered in an adult branch, have more or less completely lost their assimilative function, and merely serve as a protection to the bud. The hypopod plays an important part in this modification, but there are evidences of a partial degeneration, the epipod being reduced in size in proportion as the assimilating function is lost.[1]

The apical leaves also have partially lost their assimilative function, and have assumed, like the basilar leaves, a protective function which they exercise on the floral buds. This modification is effected differently to that of the basilar leaves, but is equally attended with evidences of degeneration. The hypopod continues to gain, as the epipod loses, in importance, but the leaflets, instead of de-

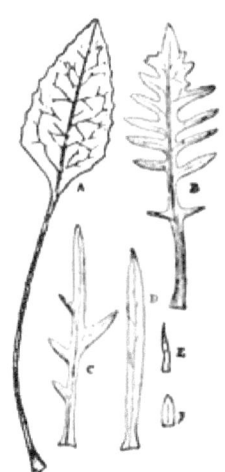

FIG. 39.—Leaves from the same branch of *Serratula centauroides*. A, foliage leaf ; B, C, D, E, successive apical leaves; F, bract of the involucre.

---

[1] On examining the foliage leaves of a plant from above downwards, one finds successively leaves with large stipules and small leaflets (fig 38, E), then leaves with very small leaflets crowded at the end of the hypopod and which have only the basilar part of the petiole (fig. 38, D), and then leaves of which the epipod and the free part of the petiole have disappeared (fig. 38, C). Finally, at the base of the plant there are leaves in which the hypopod is markedly reduced, and which carry nothing but small stipules (fig. 38, A and B).

creasing in size, decrease in number. At first there may be three leaflets, and then only a single large leaflet (fig. 38, G). Finally, towards the floral end of the branch the leaflet becomes smaller and smaller until it disappears altogether (fig. 38, H and I). The apical leaves, like the basilar leaves, are eventually reduced simply to the hypopod.

2. *Serratula centauroides.*—The foliage leaves of this species are close to the ground, their bases are ensheathed, and the long petioles terminate in rhomboidal blades.

In passing on from these assimilative leaves to the basilar and apical leaves, a double adaptation becomes apparent, as in *Rosa rugosa*. In *Rosa rugosa*, however, both adaptations were protective, while in *Serratula centauroides* this

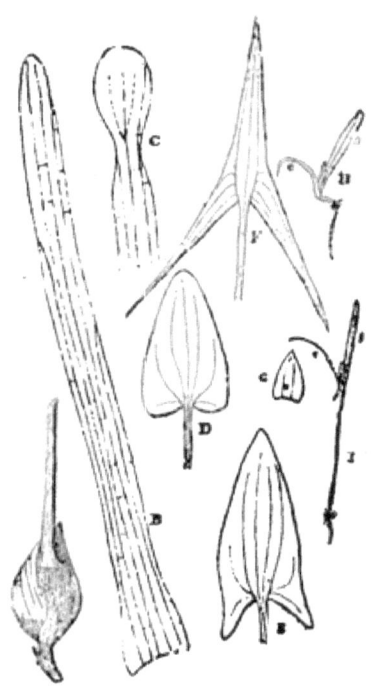

FIG. 40.—*Sagittaria sagittifolia.*
A, winter bud with membranous leaves enclosing the swollen portion and ro led leaves protecting the bud; B,C, submerged leaves; D,E, floating leaves; F, aërial foliage leaf; G, leaf protecting the flower-bud; H, seedling grown from a seed planted in the water on the surface of the mud. I, seedling from seed planted in the water under a layer of mud. In figures H and I, *c* is the cotyledon, *l* the first leaves.

is only the case with the basilar leaves, which surround the winter buds.

The modifications of the basilar and apical leaves of *Serratula centauroides* are similar to those of *Rosa rugosa*, and it is therefore unnecessary to repeat the description.

Those of the apical leaves tend towards their protection from rain. When rain falls upon the radical leaves, they are flattened out upon the ground, which prevents them from being torn. This is not so with the apical leaves, and had their primitive structure remained unchanged, they would have been constantly exposed to destruction. M. Stahl[1] has shown how these leaves have protected themselves by shortening their petioles and fringing their edges. In this way the leaf has acquired greater resistance, and the water is better enabled to run off it. This modification, however, although one advantageous to the plant, has involved the partial degeneration of the petiole and the blade.[2]

3. *Sagittaria Sagittifolia*.—The Sagittaria is an

[1] E. Stahl, *Regenfall und Blattgestalt* (*Ann. du Jard. bot. de Buitenzorg*, vol. xi., 1893).

[2] In the leaves (fig. 39, B and C), immediately above the foliage leaves, the petiole gradually disappears altogether, while the blade has become deeply serrated, thus providing for the water drainage while retaining an assimilative surface.

Further up the stem the leaf blades are more and more reduced, until eventually they are no longer differentiated into hypopod and epipod, but arise directly from the growth of the primordial leaf (fig. 39, D to F).

aquatic plant, the foliage leaves of which emerge from the water. The blade of these leaves (fig. 40, F) is shaped like an arrow-head, and the barbs facilitate rain-water drainage. The petiole is very long, and terminates in a sheathed base derived from the hypopod.

The apical leaves (fig. 4, G) are represented by transparent membraneous scales, which serve as a protection to the floral buds.

The basilar leaves, which either float on the surface of the water or else are entirely submerged, have undergone both development and degeneration. Degeneration is more noticeable on glancing downwards from the foliage leaves to those which protect the winter buds.[1]

[1] (a) The leaves nearest to the foliage leaves float upon the surface of the water (fig. 40, E). The lateral barbs, being no longer required, have begun to degenerate.

(b) The floating leaves immediately beneath exhibit barbs which are still more reduced (fig. 40, D), and there are no stomata except those upon the upper surface.

(c) The next leaves are completely submerged and lengthened out into long ribbons which broaden into flat blades at their upper ends, which are not separated from the hypopod by petioles (fig. 40, c). In this case the blade is very much smaller, and the petiole has disappeared.

(d) The next leaves to these are submerged, and are ribbon-like in shape. They spring immediately from the primordial leaf without differentiating into hypopod, mesopod, and epipod.

(e) After these leaves follow leaves consisting solely of a hypopod, which is much reduced in size (fig. 40, A). These leaves are folded round, and serve as a protection to the bud.

Lathyrus Aphaca.—The foliage leaf (fig. 4, leaves 4 and 5), which most nearly resembles the primitive type,[1] consists of two stipules and a petiole which terminates in a point and is provided with a pair of lateral leaflets.

Each individual plant is furnished with from one to three of these foliage leaves.

Above them are leaves (fig. 41, leaf 6) consisting of only a hypopod which forms two large stipules that partially enclose a small point arising from the epipod. The entire function of assimilation is exercised by these stipules, and consequently both

> (*f*) Finally, there are some scaly leaves which also consist merely of a hypopod. These protect the tubers (fig. 40, A), and are practically devoid of chlorophyll. These various leaves follow one another in the course of the growth of the plant, in the inverse order to that in which they are here described. As winter approaches, some upright stems appear, the inflated ends of which are stored with reserve nutrition (fig. 40, A). It is round these tubers that the scaly leaves are to be found. The bud which does not develop until the following Spring is at the top, surrounded by folded leaves. The ribbon-like leaves which grow under water and also those broadened out at the ends make their appearance in the course of the Summer. Next come the floating leaves with slightly developed barbs, then the leaves which rise well above the surface of the water, and finally the scaly leaves which grow out of the flowering stalk.

[1] The foliage leaves of *Lathyrus Aphaca* may be regarded as the best representatives of the primitive leaf, for in most species of the genus *Lathyrus*, each leaf is furnished with leaves and with one or more tendrils. *L. Aphaca* is probably specialized from other species, for the majority of the leaves have lost their lateral leaflets.

the petiole and the lateral leaflets have atrophied. Higher up on the stem (fig. 41, leaves 7 and 8) are some leaves consisting of two stipules and a non-branching tendril. The latter represents the mesopod and the epipod from which the lateral leaflets have disappeared. Below the foliage leaves, there are usually three leaves which have considerably degenerated (fig. 41, leaves 1 to 3). The hypopod is represented by two very small stipules, and the epipod by the tiny point in the centre. The portion of the stem out of which the basilar leaves would grow does not, under normal conditions, emerge from the soil, which accounts for the way in which these leaves have degenerated.

FIG. 41.—Seedling of *Lathyrus Aphaca*.
1, 2, 3, very rudimentary leaves; 4, 5, foliage leaves; 6, leaf composed only of a pair of stipules. 7, 8, terminal leaves turned into tendrils.

5. *Nymphaea dentata.*—Here the foliage leaves are large, floating, fleshy,

and toothed at the edges. During germination the seed-leaves remain within the seed as with *Lathyrus Aphaca*.

The first leaf is upright and reed-like in shape (fig. 42, A 1). Its principal function is to pierce the layer of soil which covers the grain, thus allowing the little terminal bud to emerge into the light. Compared with the other leaves, this needle-like leaf seems to have considerably degenerated. The plant next produces a series of little slender short-stemmed leaves which grow under the water. The first of these leaves are narrow and ribbon-like (fig. 42, A 2); the next are broader (fig. 42, B), then follow more slender submerged leaves, but with no lateral barbs (fig. 42, C); then floating leaves, the edges of which are not dentated, and last of all there are some leaves with dentated edges. Owing to their submergence, the basilar leaves of *Nymphaea* have greatly degener-

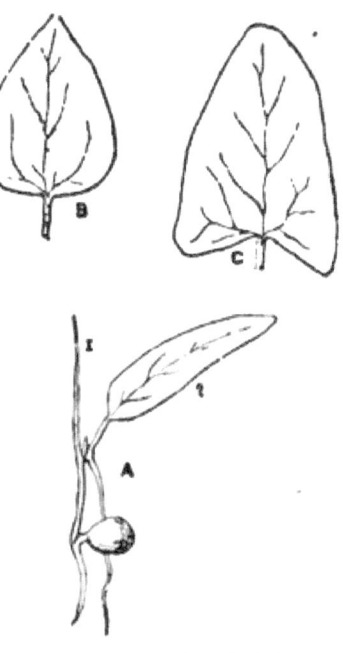

Fig. 42.—*Nymphaea dentata*.
A, seedling with two leaves (1, 2); B, C, leaves taken from older plants.

ated; the blade is very thin, being composed of a few layers of cells, and there are no stomata.

The basilar leaves and the apical leaves frequently lose either partially or entirely the assimilative function which they previously exercised. Some of these leaves serve as a protection to the buds of both leaves and flowers. Others are adapted as protection from rain, and others have undergone considerable modifications owing to their existence either under water or under ground.

In all these instances of modification it can be seen that some degeneration has invariably attended each change which has taken place.

§ 5.—*Modification of organs which are homologous in individuals of different species.*

Having investigated the modifications of leaves in the individual, we will now give our exclusive attention to the various modifications undergone by the foliage leaves; we shall see among the various species we examine that degeneration has played a part in each instance of modification.

Of these adaptations the following are the most characteristic: adaptation to climbing, to a carnivorous diet, to aquatic life, to defence against ants, against drought, and against herbivorous animals.

1. *Adaptation to climbing.* — Climbing plants attach themselves by means of tendrils to the nearest support within their reach. These tendrils,

which are thread-like and sensitive to contact, are

Fig. 43.—*Cobaea scandens.*
A, B, seedlings; the leaves of the second end in tendrils; C, young leaf of adult plant.

modifications of either stems, leaves, or roots, the latter being the most rare. We will confine our attention at present to those adapted from leaves.

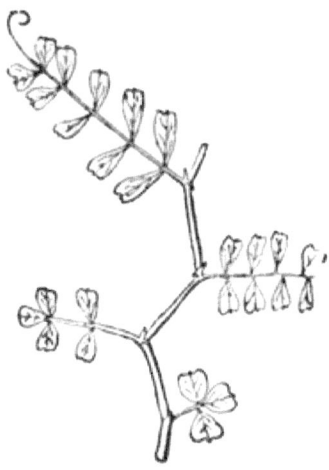

In *Cobaea scandens* the first two leaves of the seedling usually terminate in a leaflet (fig. 43, A); the leaves which come after are much finer and are modified into tendrils (fig. 43, C).

This modification of leaflets into tendrils, which entails the almost com-

Fig. 44.—Basal portion of a young plant of *Vicia Pyrenaica.*

plete disappearance of the original assimilative function of the leaflet, is by no means uncommon. *Vicia Pyrenaica* (fig. 44) furnishes a good example of this change, of which the degeneration of the blade is the necessary consequence.

In *Cucumis sativus* (the cucumber) some of the leaves are the shape of ordinary assimilative leaves, while others are entirely modified into tendrils; in the case of the latter the blade has completely degenerated.

2. *Adaptation to carnivorous nutrition.*—Some plants, instead of obtaining nutriment exclusively from minerals and carbon dioxide, are capable of assimilating animal matter from insects and other small organisms, which they capture by means of a special function exercised by the leaves.

(*a*) *Utricularia.*—In the aquatic *Utricularia* the leaves are minutely sub-divided like many other submerged plants. The leaves of *Utricularia vulgaris* exhibit leaves of this type. Some of their ramifications carry pouches, which serve for the capture of minute organisms. These pouches are formed by the modification of part of the blade, and have almost completely lost their chlorophyll.

Utricularia intermedia (fig. 45) exhibits two kinds of branches; the one kind stretches out horizontally and carries green leaves; the others are devoid of chlorophyll, and the leaves are purely carnivorous.

(*b*) *Nepenthe.*—The distal extremity of the leaf

MODIFICATION OF THE ORGANS OF PLANTS

terminates in a receptacle of complicated structure which is adapted for the capture of carnivorous nutriment. Degeneration has attended this change, for the modified portion of the leaf is almost

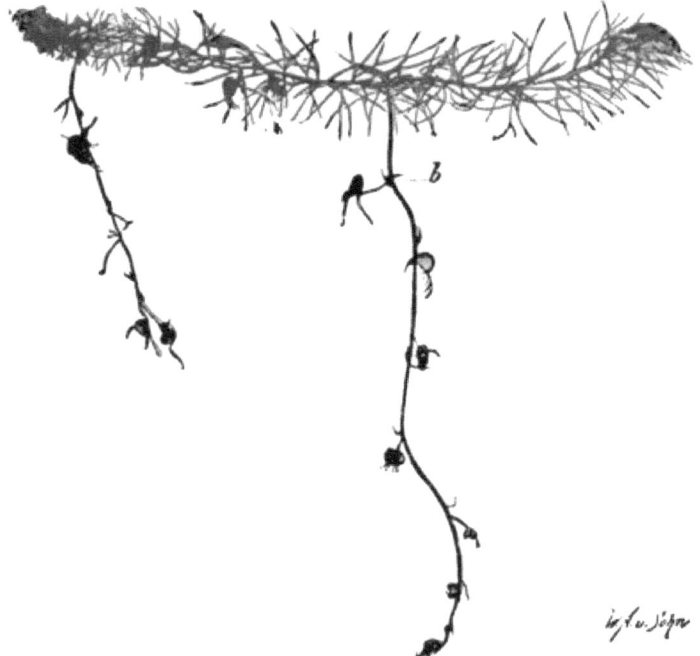

FIG. 45.—*Utricularia intermedia* (after Goebel, *Pflanzenbiologische Schilderungen*, vol. II., p. 135).
The plant has grown from a winter bud of which the remains are visible to the left. The horizontal, assimilating branch bears two pendulous twigs with carnivorous urns.

entirely without chlorophyll. A carnivorous diet, however, for these plants, is only supplementary to the typical plant nutrition, and therefore the assimilative function remains active.

F

In *Utricularia* this function is exercised by the leaves (as in *U. intermedia*), or by certain parts of the leaves (*U. vulgaris*). In *Nepenthe*, however, another and different assimilative organ is exhibited. This consists of two lateral herbaceous growths, which are not formed from the blade, but arise from the petiole. In between the receptacle and the enlarged portion of the petiole is a part in which the petiole has assumed the function of a tendril. While the receptacle is in process of formation, this tendril twines itself round a support in order to obtain the additional strength that will be required later on to uphold the receptacle when full of digestive juice.

(c) *Drosera*.—The leaves of the sundews (*Drosera rotundifolia, D. longifolia*, etc.) are furnished with a great number of emergences, each of which terminates in a digestive gland. There is a large drop of sticky fluid at the end of each gland, and these drops, which sparkle in the sunlight, have given the plant its name (*Ros solis* = sundew) (fig. 46).

These emergences contain very little chlorophyll.

"A plant of *Drosera*, with the edges of its leaves curled inwards so as to form a temporary stomach, with the glands of the closely-inflected tentacles pouring forth their acid secretion, which dissolves animal matter afterwards to be absorbed, may be said to feed like an animal."
—Darwin, *Insectivorous Plants*, p. 18.

This adaptation to a carnivorous diet has not

come about without a certain amount of consequent degeneration. The chlorophyll has disappeared, excepting a small quantity exhibited in the upper and under surfaces of the blades of the leaves, in the flower stalks, in the central tentacles, and in the petioles.

3. *Adaptation to an aquatic life.*—The adaptation of leaves, particularly of submerged leaves, to an aquatic life entails important changes, and these changes are invariably attended by degeneration.

1. Owing to the buoyancy of water, submerged leaves are not dependent on any highly developed organic support, and the fibres and other thick-walled cells have consequently degenerated to a great extent.

2. There being no transpiration in submerged leaves and very little in floating leaves, the

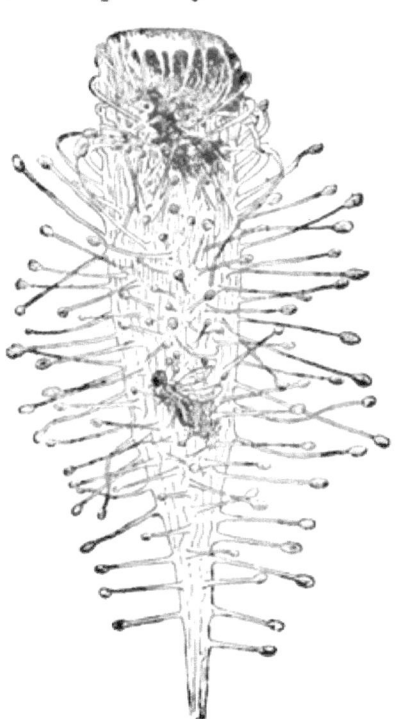

FIG. 46.—Leaf of *Drosera longifolia*. Near the top, an insect is in the act of being captured and the suckers are bending over it. Lower down is an insect reduced to a chitonous skeleton.

former are very seldom provided with stomata, and the latter only exhibit stomata on their upper surfaces (see page 26, note on *Ranunculus and Stratiotes*). The conducting apparatus too, *i.e.* the wood and the roots, has greatly diminished in size owing to the diminution of transpiration.

3. Very little light penetrates to submerged leaves, and oxygen is only partially soluble in water. The blade has therefore become much smaller. Very little importance, however, is attached to this last modification, as the size of the blade varies greatly in different specimens, whereas the other modifications are similar in all.

(*a*) In the species already mentioned—*Sagittaria* (p. 72) and *Nymphaea* (p. 76)—the first leaves which are formed are submerged, with petioles either short or entirely absent, and very slender blades. In *Vallisneria* and many species of *Potamogeton* there are only slender submerged leaves.

All these leaves augment their surface contact with the water by ridding themselves of the deeper layers of assimilative cells. Very little light could penetrate to these cells, and any oxygen formed would be difficult to get rid of.

(*b*) In other plants, especially in the Dicotyledons, another kind of modification takes place for the same end; the blade is divided into very thin segments which in some cases present a hair-like appearance.

The *Ranunculus* is a good example of this.

Besides the many species found on moors and cultivated land and in woods, there are other kinds of *Ranunculus*—*R. sceleratus*, for instance—which grow by the water-side. In this species it is not uncommon to find floating leaves, and leaves which wholly emerge from the water, growing on the same stem. In other species, such as *Ranunculus hederaceus*, there are seldom any but floating leaves. Among the numerous varieties of *Ranunculus aquatilis* there are some which, in addition to the floating leaves, have also some leaves which are completely submerged and deeply dentated; in other varieties of the same species there are only slightly dentated leaves, while in *Ranunculus fluitans* the leaves are all fringed.[1]

(c) In a submerged plant which is to be found in Madagascar, *Ouvirandra fenestralis* (fig. 47), the diminution in volume, compared to the surface, is effected in a different manner. After the leaf has developed in the ordinary way, all the parts of the leaf in between the veins of the blade disappear. This results in the blade being reduced to a delicate piece of network, consisting entirely of communicating veins accompanied by adjacent assimilative cells.

4. *Adaptation to defence against ants.*—In the

---

[1] *Ranunculus sceleratus* grows by the water-side; *R. hederaceus* is to be found on marshy ground, the internodes and petioles being quite short; *R. aquatilis* grows in shallow water, and the internodes are of sufficient length to raise the flowers out of the water; *R. fluitans* grows in strongly-running streams, so that floating leaves would be useless.

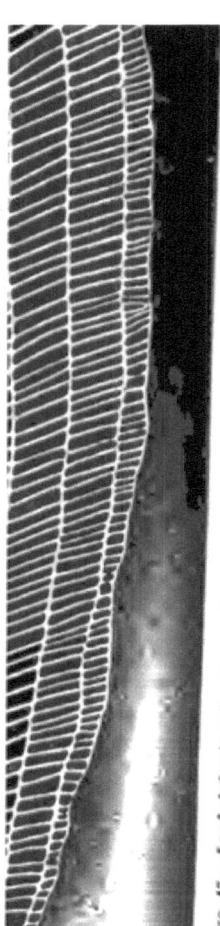

FIG. 47.—Leaf of *Ouvirandra fenestralis* (after Goebel, *Pflanzenbiologische Schilderungen*, vol. II., p. 320). The print was taken by a direct impression from the leaf.

tropics, especially in South America, plants are much exposed to the ravages of leaf-eating ants (Atta). In order to protect themselves, they provide other kinds of ants with shelter and nourishment, and this necessarily entails several modifications.

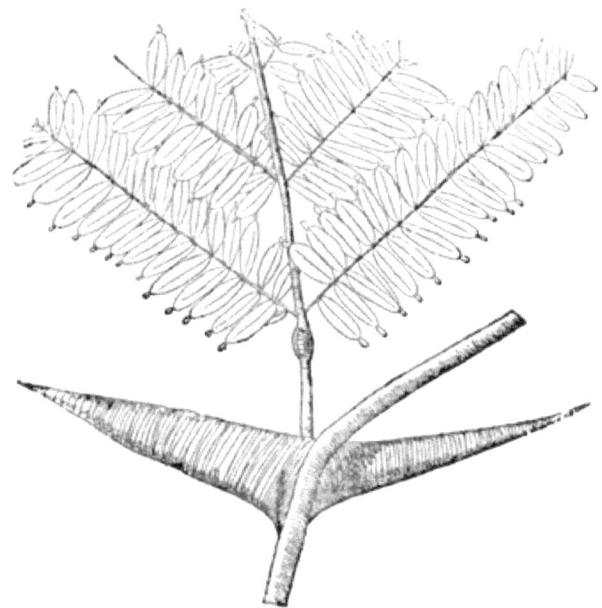

FIG. 48.—Part of a leaf of *Acacia sphaerocephala* (after Schimper, *Die Wechselbeziehungen zwischen Pflanzen und Ameisen im tropischen Amerika*. Fasc. 1 of *Schimper's botanische Mittheilungen aus den Tropen*. Jena, 1888).

In *Acacia sphaerocephala* (fig. 48), for instance, the blades of the leaves are bipinnated and the proximal leaflets terminate in small cavities filled with a nutritive secretion. The stipules, which no longer contain chlorophyll, are modified into hollow thorns in which the ants live and obtain nutriment, not only at the ends of the leaflets, but also in a thorny gland which is situated upon the petiole.

5. *Adaptation to drought.* — In places where rain seldom falls, plants are provided with natural reservoirs of water. These reservoirs are situated either in the roots or the stems and occasionally in the leaves. Where the leaves are fleshy, these reservoir leaves are generally very simple in formation. Those of *Sempervivum* and of several other genera arise directly from the primordial, non-differentiated leaf, and there is no differentiation into hypopod, epipod and petiole (fig. 49). See further on (fig. 73), p. 236 (*Sempervivum tectorum*).

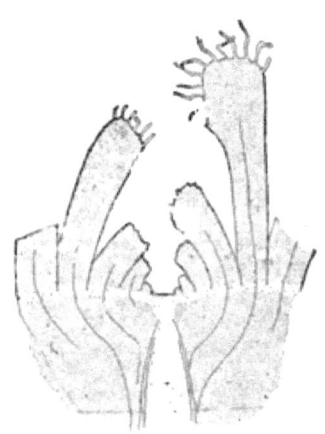

FIG. 49.—Growing point of *Sempervivum arachnoideum*.

FIG. 50.—Branch of *Caragana*.

6. *Adaptation to defence against herbivorous animals.*—Thorns are an adaptation which serve as a protection against herbivorous animals. They

MODIFICATION OF THE ORGANS OF PLANTS   89

are derived from the modifications of various parts of the plant, either the roots, stems, leaves, or even of the floral stems.

This modification is always accompanied by some degeneration; the thorn, which is quite hard, ex-

FIG. 51.—*Mamillaria elephantidens* (after Lemaire? The figure is taken from Goebel, *Pflanzenbiologische Schilderungen*, vol. l., p. 71).

cepting at the point, is made up of thick-walled cells from which the protoplasm has disappeared.

Leaf thorns arise from the modification of various parts of the leaves—either of the stipules (as in *Caragana*), the petioles (*id.*) (fig. 50), or of the blade (as in *Ilex*). In each case the leaf partially retains its assimilating function and its chlorophyll.

In other plants, however, and especially in Cacti, and the fleshy-leaved Euphorbias, the leaves exhibit further evidences of degeneration, their function being exclusively one of defence (fig. 51).

CHAPTER II

IN THE EVOLUTION OF INSTITUTIONS ALL MODIFICATION IS NECESSARILY ACCOMPANIED BY DEGENERATION

THE distinction we have drawn between the homodynamic organs of an organism, and the homologous organs of organisms belonging to other species, is not applicable in sociology, as we have already pointed out in the introduction. Institutions, however, may be regarded from two distinct standpoints—from a statical point of view, as they exist in the same society, and from the dynamical point of view, as existing from epoch to epoch, and from society to society. In both cases, we shall arrive at the same conclusion as in biology, that all modification entails degeneration.

In order to demonstrate this we will examine in succession the modifications undergone by the principal types of financial organizations now existing in Europe, and the most important stages in the evolution of landed property amongst various peoples.

§ 1.—*Modifications of similar institutions in the same society.*

The financial organization of European towns and states has undergone very important changes since the middle ages. Taxes and duties have attained a now universal importance, as substitution for the revenue from crown lands, which constituted the principal, if not the only, resources of the sovereigns of the feudal ages.[1]

---

[1] There are three stages in the evolution of financial systems in countries (such as England, for instance) where the question of finance has been most successfully dealt with.
  1. *The Feudal System*, wherein the king had no separate revenue apart from the nation, and wherein the revenue of the sovereign was principally derived from crown land, the cultivation and administration of which was carried on as a source of private income to the king.
  2. *The Co-existence of the Feudal System and the Modern System*, viz., the disappearance of personally held land and its attendant institutions, the development of the public property of the State or Township, and the imposition of duties and taxes.
  3. *The Modern System*, viz., the complete separation of the personal property of the sovereign from the property of the nation, the increasing importance of taxes and duties, and the almost complete disappearance of State and town lands. Industries taken over by the State—such as railways, postal and telegraphic arrangements, etc.—and by the towns—such as gas, water, etc.—constitute monopolies, and are no longer subjected to the law of competition which is always active in private business. — WAGNER.

Here, too, we find that all modification is attended by degeneration. As a demonstration of this, we will examine in succession—firstly, the Communal budgets of Belgium; secondly, the State budgets of the States which compose the German Empire; thirdly, the budgets of Germany, England, and France, regarded as belonging to one group—the Western Republic of Auguste Comte.

I. *The Communal Budgets of Belgium.*

Collective property in its archaic form still exists, in spite of considerable modifications in certain parts of the Ardennes, of the Fagne, and of Lower Luxembourg.[1]

At Wanlin, for instance, the collective property constitutes nearly one-half of the land—220 hectares, of which 130 is arable land, and 90 is woodland. The arable land is divided into allotments, apportioned among the heads of the various families every eighteen years, for which they pay an annual rent of 10 francs. There are no taxes or communal duties, the revenue derived from this public property being sufficient without them.

This economic and fiscal system, suitable to a scattered and agricultural people, existed formerly throughout the whole country, but the increase in population and cultivation of the land have almost

[1] Paul Errera, *Les Masuirs*. Historical and judicial researches in the vestiges of the old territorial system of Belgium. Brussels, Weissenbruch, 1891.

put an end to this archaic condition of things. Private property has encroached upon collective property, and taxes and duties are now substituted for the revenue which was formerly derived from land alone.

This modification was of a progressive nature, the extended requirements of the population demanding a corresponding augmentation of the communal revenue. Degeneration, however, is exhibited in the disappearance of the collective land of the communes and of the attendant administrative institutions.

The monographs of Paul Errera and the documents collected in 1875 by the Statistical Commission relating to the expenditure and receipts of the communes, demonstrate the various stages of this degenerative evolution:—

1. The vestiges of the old system may be observed to decrease in number and importance on approaching the North-West from the South. The proportion of receipts from land compared with those derived from taxes and duties is greatest in Upper Belgium; then follow those of Condroz, Hainault, Brabant, Campine, and finally, of the two Flanders.
2. The constitution of the two Flanders passed through the following stages:—

(a) In certain localities round about Bruges the old and the new system co-exist. Side by side, for instance, with the communes of Oedelem, Beernem and Oostcamp we find, though in a degenerate condition, the old collective property of the *aenborgers* of *Beverhoutsveld*, with its administrative college of *Veldheeren*.[1]

(*b*) In Sysseele, and in several other communes, the assembly of Veldheeren has been abolished, and the property which they administrated has become the property of the commune by whom the rents are collected.

(*c*) At Gand, the grazing land of the *Heernisse* of Saint-Bavon, which was declared collective property by a decree of November 16, 1887, was afterwards

---

[1] Beverhoutsveld is a vast domain situated outside the gates of Bruges, in the territory of the commune of Oedelem. According to the terms of a decision of the 13th August 1859, it constitutes a collective property belonging to three sections of the communes of Oedelem, of Beernem, and of Oostcamp. The family representatives of these three sections are alone privileged to rent these lands, and the revenue accruing from them is divided among the three sections.

The old administrative authority, the college of Veldheeren, which was in existence in the thirteenth century, still exists, though in a new form. According to the terms of the decision of 1859, the police are provided for from the taxes of Oedelem, as the property is within that district; "but in other matters, and in the administration of the rights and revenues accruing from it, the property shall be under the control of a commissioner appointed, so far as possible, by the interested parties—*i.e.* the aenborgers—in conformity with the laws and regulations relating to the matter."

The Vrij-Geweid, also situated in the environs of Bruges, is similarly administrated. (See Errera, *des Masuirs*, chaps. xix., xx. and following.)

transformed into private estates, and has been so built over as to become almost unrecognizable.

(d) In some towns the receipts from land only represent a very small part of the communal revenue, 3 per cent. at Roulers, 0·1 at Saint-Nicolas.[1]

It is thus evident that the evolution of communal finance, which has been especially characterized by the development of duties and taxes, has been accompanied by the degeneration of the old system of collective property.

II. *Budget of the States of the German Empire.*

The feudal system of finance has left its mark upon Germany, especially in the eighteen subsidiary states where archaic institutions have been best preserved.[2] The principal stages in the progressive evolution of the modern system, and the corresponding degeneration of the feudal system may be enumerated as follows:—

1. Mecklembourg-Strelitz still exhibits the primitive system in a comparatively little altered form. The Grand Duke is the sole administrator; the budget is not separated from the civil list; the revenue amounts to an average

---

[1] Hector Denis, *l'Impôt*, pp. 55. Brussels, V⁰. Monnom, 1889.

[2] Leroy-Beaulieu, *Traité de la science des Finances*, 1, 21, and following.

Adolf Wagner, *Lehr-und Handbuch der Politischen Oekonomie; Finanzwissenschaft.* Erster Theil, secs. 214-216.

of seven millions, of which five are derived from land.

2. The Duchy of Mecklembourg-Schwerin is administered partly on the modern and partly on the feudal system, but the former predominates. There are two public budgets. That of the Duke is of the greater importance, and amounted to 14,500,000 marks in 1887, of which 7,000,000 were derived from landed property.[1] The second budget is that of the municipal expenditure (*Gemeinsamen Finanzverwaltung*), which is defrayed by means of rates and taxes, the receipts of which amounted in 1887 to 4,175,000 marks.

3. In the larger States of the Empire the feudal system has completely disappeared, but the revenue obtained from collective property forms an important part of the public income.

Bavaria	17·3 %
Würtemberg	13·2 %
Saxony	9·7 %
Baden	7·1 %
Prussia	8·4 %

4. In the budget of the Empire, which is of recent construction, naturally no traces of the feudal system remain.

---

[1] These lands comprise more than 99 German square miles with over 206,000 inhabitants—about 37 per cent. of the population. L. Beaulieu, i., p. 43.

III. *The Budgets of Germany, France, and England.*

(a) *Germany.*—We have just seen that in the principal States of the Empire there is still an extensive amount of collective property. In Prussia, for instance, besides forest land, there are nearly 1,500,000 acres of arable land, an area equal in extent to one of the smaller departments of France.

(b) *France.*—Of the old collective property of France only forest land remains, all grazing and arable land having long since been alienated.

The budgets of modern times, however, still exhibit traces of the old feudal system. The State, for instance, up to the last few years, continued to receive quit-rents for properties under the old system.[1] The budget shows an annual decrease in these receipts, having fallen from 100,000 francs in 1857 to 32 francs in 1869. In 1876, however, there was a rise to 2000 francs.

(c) *England.*—Here the decline of collective property is still further exhibited. In the Statistical Abstract published in 1877, the net revenue from Crown lands for Great Britain and Ireland figures at £40,000 net. Among the "miscellaneous receipts" we find only £200,000 which can be regarded as revenue derived from collective property, *i.e.* 15 millions in a budget

---

[1] Leroy-Beaulieu, *Traité de la science des finances*, i., p. 35.

receipt of 1,956,000,000 francs. Later on, in 1884-1885, these revenues became rather larger (principally owing to the purchase of the shares in the Suez Canal which had belonged to the Viceroy of Egypt), but the proportion remains the same—22 to 23 millions in a budget of 2 milliards 300 millions. We see, then, that in all the financial systems of modern Europe the progressive evolution of the modern system is taking the place of the more or less rapidly-degenerating feudal system.

§ 2. *Modification of similar institutions in different social groups.*

According to the primitive constitution of things, the land occupied by a tribe or clan was regarded as *res nullius*, and consequently at the free disposal of all the members of the community (the *Feld- Walt- und Weidegemeinschaft* of V. Maurer).

With the increase of population, the value of land rose, and the state of things became modified, the rights of groups and individuals becoming consolidated and at the same time limited. Then arose gradually or simultaneously the following various forms of landed property:—(1) Land held by families; (2) by villages; (3) feudal property; (4) communal or public property; (5) property belonging to corporations; (6) private property.

Family, village, and feudal property represent,

among certain peoples, three successive stages in the evolution of property. When the old system of land tenure was abolished, private and communal or public property began to develop simultaneously.

While certain lands which were free to all the inhabitants became transformed into collective property, other such lands lost their public character and became private property. In the first case, the communes, on being called upon to fulfil functions of increasing complexity, proceeded to transform all or part of the properties concerned into patrimonial property or property for the use of the people (*communaux, allmenden*).[1]

In the second case the property of the old community became the joint but undivided property of the members of the corporation; when, however, for purposes of cultivation it became necessary to divide it, the corporative property became transformed into private property.

---

[1] Giron, *Le droit administratif de la Belgique*, No. 683. "There were three kinds of communal property:

"(a) Property directly appropriated to the use of the public such as public squares, streets, churches, &c.

"(b) *Communal* property properly speaking—*i.e.* the real estate and rights belonging to the tribune and to which the people were entitled to a personal share. These consisted of the forest land, rights of appanage, waste land, moorland, and the rights of pasturing.

"(c) Patrimonial property, *i.e.* that held by the commune, the revenue from which went to the commune to defray the expenses of administration. It included timber land, arable land, house property, market places, &c."

On reviewing in succession these various phases in the evolution of landed property, it will clearly be seen that modification has everywhere been attended by degeneration.

1. *Family property (Montenegro).*—Of all the Balkan States, Montenegro—owing to the natural barriers which separate it from the rest of Europe—has best preserved its archaic institutions. Here, side by side with modern institutions, may be found the old system of division into forty-two tribes (pleme) which are sub-divided into clans or confraternities (brastvo) and into patriarchal families (zadrugas and inokosnas).[1] The development of modern political and judicial institutions has, however, considerably lessened the importance of the plemes and the brastvos, so that progression in this direction has not been effected without accompanying degeneration.

With regard to property, the two different forms of family tenure have been substituted for what was formerly the tribal or clan system. Of the former collective property of the clan, there only remain the following traces :—

    1. Property rights held over certain portions of land—generally forest or waste land.

    2. The right of pre-emption in favour of members of the brastvo or of those related to

---

[1] For information about the common or differential characters of the zadruga and the inokosna see Ardent, *La Famille zougoslave au Monténégro*. (*Réforme sociale*, 17th October 1888.)

a member within the first six degrees of lineal descent.[1]

3. The right of allotting to relatives their share in the duties of helping widows and paupers in their work. The workers in this case receive no payment, neither have they any right to demand maintenance.

Still rarer are vestiges of the collective property of the pleme. A few portions of land, however, still belong to that body, and it is probably a survival of this ancient condition of things that foreigners are

---

[1] Article 48 of the Civil Code of 1888, drawn up by Bogisic in all possible accordance with "the excellent customs" of Montenegro, begins with the statement that "the right of pre-emption, a privilege which has so long been enjoyed by the members of the brastvo, by persons whose lands adjoin, and by the members of the village and pleme, still flourishes, and will probably continue to do so."

Bogisic adds that, in accordance with this right, "any person desiring to sell his land, or any kind of real estate belonging to him, is constrained, according to the established custom in such cases, to first offer it in legal order to those persons who enjoy the right of pre-emption, in order to give them an opportunity of purchasing it at the price at which it is to be offered to the public.

Article 49, sec. 1, gives a list setting out the order of precedence of those who enjoy the right of pre-emption.

1. Members of the brastvo within the first six lineal degrees of descent.
2. Persons owning adjoining lands.
3. The other members of the village.
4. The other members of the pleme.

} Transference, of recent origin, to neighbours of rights originally confined to relatives.

If none of those entitled to the first offer desire to purchase, the owner may then sell his property to any other Montenegrin.

not permitted to acquire landed property in Montenegro,[1] and that the public are unrestricted in the right to hunt over any ground they choose.[2]

2. *Village property (Russia).*—Village communes and the periodical division of land—the *mer* of Russia or the *dessah* of Java, for instance—do not represent types of a primitive system, but are the outcome of a whole series of modifications. Kowalevsky traces the evolution of the present system in Russia through the following principal stages:—

    1. The joint use of land by the members of one family group (*pechische*), corresponding to the *zadruga* of the Southern Servians, and sometimes comprised of more than forty persons.

    2. The division of the mother-family into separate households, thus forming a village community, and the temporary allotment of the land of the community among the separate families.

---

[1] The old Montenegrin law relating to landed property, which prescribes the purchase of land in Montenegro by any but Montenegrins, is still in full force. No transaction in violation of it is legally binding (Dickel, *Étude sur le nouveau Code civil monténégrin*).

[2] Throughout the Southern Slavonic countries hunting is the free right of all. Anyone may hunt where and how they please, not only on public ground, among the mountains and forests, but upon private property, whether cultivated or not (Dickel, p. 36).

3. The alienation of all or part of the allotments assigned to the families and the constitution of agricultural communes no longer necessarily consisting exclusively of persons akin.
4. The periodical division of land, which, as the population increased, was instituted with a view to re-establishing an equal distribution. This system of division, which was established gradually, only dealt with the more valuable sort of land, such as meadows and arable land. The forest land and pasturage, that at least which was not already annexed by the Lords of the Manor, was free to all.

This transformation of family communities into village communities was not effected without accompanying degeneration. The administrative institutions of the family group disappeared, and the rights of pre-emption in favour of blood relations were gradually replaced by village rights. The importance of the family, regarded as an economic group, decreased in proportion with the increase of the importance of the village.

In some places, however, and especially among the Ossetes who inhabit the valleys of the Caucasus, the old system may still be found. There, at any rate up to within the last few years, the *aouls* (villages) are principally comprised of families holding land in joint tenure, frequently sharing all things in common. These *aouls* are very rarely met with nowadays.[1]

Besides these family communities, there yet remain among the Ossetes, as in Montenegro, numerous vestiges of the primitive system of clan property, *i.e.* the appropriation of certain portions of land by the members of the clan, the common use of pasturage and forest land, the enforced participation in certain public works, and the rights of heritage over unclaimed land, or unappropriated property which had become so owing to the lapse of some "feu" or by the extinction of a family community.[2]

3. *Feudal property (England).*—The introduction of the feudal system into England resulted in the substitution of a new method of grouping, in place of the old agrarian communities. These new groups, like the *townships* of earlier times, consisted of a complete organization occupying definite boundaries. Instead, however, of being a family group administrated by a democratic organization and government, this new system, which *Sumner-Maine* calls a "manorial group," consisted of a tenantry autocratically grouped together and governed by a feudal chief—the Lord or Seignior.[3]

---

[1] See Kowalevsky, *Coutume contemporaine et loi ancienne*, p. 42. Paris, Larose, 1893.

[2] See Kowalevsky, *Coutume contemporaine et loi ancienne*, pp. 68 and following. Paris, Larose, 1893.

[3] See Sumner-Maine, *Les Communautés de village; 1 D., Études sur l'Histoire du Droit; 1 D., l'Ancien Droit et la coutume ancienne.*

Although it cannot definitely be asserted that each one of these manorial groups was developed from what had formerly been a village community, yet it is evident that such was frequently the case, and that the transformation was accompanied by partial degeneration. For instance:—

1. The assembly of the inhabitants of the township, which formerly exercised complete control over all legal and administrative affairs, disappeared, and in its place sprang up the Manorial Court which was comprised of a limited number of tenants and presided over by the Lord of the Manor or his representative.
2. The collective property became absorbed, or was at least considerably reduced by the acquisitions of the Lord of the Manor, or by divisions effected by members of the communes. The old system of collective property held by townships did not, however, wholly disappear.

(*a*) The "rights of usage" in regard to waste land, forest and moors (such as the use of unclaimed pasturage, the cutting of timber, etc.), were still enjoyed by the old inhabitants, and even in some instances by other persons upon whom these "rights" had been conferred.

Emile Laveleye, *Les Communautés de famille et de village* (*Revue d'économie politique*, 1888, pp. 350 and following).

Vinogradoff, *Villainage in England*, Oxford and London, 1892.

(b) With regard to meadow land, sometimes the Lord of the Manor put up enclosures for his own benefit from Candlemas till Midsummer, the rights of the community being established during the remainder of the year only; sometimes it was the community who put up the enclosures, when the Lord of the Manor was only entitled to the use of the land during the intervals; sometimes pasturage was held as the joint property of the old community, or rather of their descendants the tenants; but as a rule it was regarded as more or less common property. The best meadow land was divided up into what were termed "deals" and apportioned by drawing lots.

(c) With regard to arable land, the method of appropriating and cultivating the land occupied by the tenants retained many traces of the village system of collective property. For instance: the enforced rotation of crops; the periodical division of land in certain parts of the country; the division of land into three breaks in other places; and the destruction after the harvest of the enclosures surrounding the crops, after which the land was used for the herding of cattle.

These survivals may yet be found in some districts of modern England, in spite of all the great changes in the English system of property; changes such as the disappearance of the serf and the appearance of yeomen in the course of the thirteenth to the sixteenth century, and the dispossession of the yeoman in favour of the growth of large properties in the course of the sixteenth to the eighteenth century.

4. *Public or collective property (Switzerland).*— It is easy in Switzerland to follow the course of the transformation into political communes of the old system of economic communes, whether village or manorial.

In the mountainous parts of Switzerland this transformation is still incomplete, and side by side with the modern commune may be seen the old form of collective property, though in a more or less advanced stage of degeneration.

The successive stages of this evolution may be enumerated as follows :—

1. The village communities (the *Feld-Walt-und-Weidegemeinschaft* of von Maurer).
2. The collective property of the inhabitants, whether feudal, free, or partly both (*Gemischte Gemeinde*).

    The *Feldgemeinschaft* completely disappeared after the Reformation, the collective land of the community, the joint use of which was the right of all the inhabitants, being restricted to mere waste land, forest land, and pasturage (*Allmend*).[1]

---

[1] The *Allmend*, in the primitive sense of the word, meant that part of the old collective property held in joint tenancy by a community of inhabitants or any other public body, the use of which was limited to those who had a personal title to it. This primitive meaning has changed in Switzerland—excepting in the

3. The institution of *Bürgergemeinde* — public corporations — the members of which enjoyed the sole use of the *Allmend*, and upon whom fell the costs of local administration; as the population increased, the conditions of admission into the *Bürgergemeinde* became more and more strict.
4. The creation of *Einwohnergemeinde*, or political communes, established side by side with the old system, and in many localities eventually taking its place. In these the responsibilities of the former system were assumed, and part or the whole of the collective property was appropriated. Where the old communities still survive, they have, as a rule, ceased to exercise the greater part of their original functions. In the Cantons of Berne and Saint-Gall, for instance, the old communities have delivered up the greater part of their possessions to the political communes to provide for the expenses of general administration. Their only now remaining function is the administration of the small remainder of their patrimony and the maintenance of the indigent members of the community.

Canton of Schwyz—owing to the changes in the institution itself (see Miaskowski, *Die Schweitzerische Allmend in ihrer geschichtlichen Entwickelung, von xiii. Jahrhundert bis zum Gegenwart*).

Independently of this decay of the old system of communities, the formation of political communes was attended by other phenomena of degeneration:—

1. The suppression of all or part of the "rights of usage" enjoyed by the inhabitants.

The *Einwohnergemeinde*, being called upon to discharge more and more onerous and complex functions, were obliged to either partially or wholly transform the communal possessions, to the personal use of which the people were entitled, into property appropriated to the use of the public, either directly (*i.e.* into churches, teaching institutions, etc.) or indirectly, as a means of obtaining a revenue (*Erwerbsquelle*).

2. The decrease in collective property.

Many of the *Bürgergemeinde*, although no longer discharging public functions, retained part of their estates, which were held by the members in joint tenancy. On the other hand, as the increased population necessitated the cultivation of the *Allmend*, the original "right of usage" resulted in many instances in a transformation of the land into private property.

This transformation, however, was not always complete, and all the intermediate stages may be traced between the old collective tenure and the appropriation by individuals.

5. *Corporative property (Belgium).*—Here we will limit the sphere of our observations to Belgium, in

order to avoid repetition, similar examples being almost universally exhibited throughout Europe.[1]

Our information is obtained from the work of Paul Errera, entitled *Les Masuirs, recherches historiques et juridiques sur quelques vestiges des formes anciennes de la propriété en Belgique.*

The masuirs (the *amborgers* of Flanders) were the *mansuarii* or *mansoarii* of the Merovingian period, originally serfs, afterwards tenants and copy-holders, and finally freemen.

Their history exhibits the following stages:—

1. The feudal epoch in which the masuirs—*i.e.* all the members of the manorial group—enjoyed "rights of usage" over all waste lands, forests and pasturage adjoining their holdings. These rights appear to have been conferred by the Lord of the Manor, but they really dated from a much earlier period.
2. As the increasing population necessitated the regulation and limitation of these rights, certain conditions of property and residence were stipulated for in those seeking admission to the rights of the masuirs, and these privileged persons organized themselves into corporations which were more or less exclusive and separate from the general community.

[1] With regard to corporative property in Switzerland, see von Miaskowski, *Die schweizerische Allmend*, pp. 37 and following.

3. By degrees—by means of cantonments, purchases, prescriptive claims, &c.—these corporations absorbed the best part of the land, and became almost independent of the Lord of the Manor; as a rule, the latter gave up half of the common territory to them, and freed the surplus from all rights of usage. In the corporations of masuirs, however, there were still a few remaining vestiges of some of the institutions of the old manorial group from which they had gradually developed. The Lord of the Manor, for instance, himself being an inhabitant and a masuir, had a right to a share in the property of the community, and further, in his seigniorial capacity, certain privileges accrued to him such as "la haute fleur des bois," *i.e.* tithes and pannage (crops of acorns).

4. The Revolution put an end to all feudal rights, and removed the last remaining traces of the origin of the masuirs.

Throughout this long series of transformations, it is evident that degeneration has followed in the track of progress. Besides the disappearance of the manorial group and its attendant institutions, the rights of the masuirs may be said to have become more restricted as they became more defined and secure. In the early days, all the inhabitants enjoyed joint rights over a vast common territory, at the close of the old system this territory had become much reduced in extent, and had become the property of a more or less large group of privileged persons.

6. *Private property (Switzerland).*—After the Revolution, the communities of masuirs and other similar corporations ceased to have any recognized legal existence. Those which still survived in spite of the irregularity of their legal position, owed their existence to their insignificance. The others dispersed themselves, or were dispersed, and the property which had belonged to them was either incorporated with the communal estate, or divided up among the members of the old community.

In each of these cases the transformation was attended by degeneration, for the archaic administrative organization disappeared.

We saw in the *allmend* of Switzerland, this same divergent evolution of public and private property, part of the common land being transformed into communal property, while the use of the surplus ended in some instances in the land becoming ultimately the private property of individuals. This frequently occurred where land was cultivated as orchards. In early times both fruit and fruit-trees belonged, like the land itself, to the community, and in certain parts of the Cantons of Uri and Schwyz this is still the case. By degrees, however, individual rights over fruit-trees planted on the *allmend* came to be recognized. These rights, whether temporary or held for a life-time, eventually became perpetual, and finally this right to the private acquisition of trees led to a right to acquire the land itself. This last transformation was not effected without a struggle and occasionally the land was reclaimed by the community, the proprietor of the trees receiving compensation. Nowa-days the possession of trees and land usually go together. Duality of this kind, however, is still to be met with in certain localities. In the *Sernfthal* (in the Canton of Glaris) a still stranger custom prevails with regard to the maple forests. There, the soil, the trees, and the fallen leaves (the latter being used as litter for cattle) all belong to different persons.[1] With regard to house property there are more intermediary conditions between use and possession. In some villages, the chalets as well as the ground upon which they are built, belong to the whole community; in other villages, both are part of the collective property. Sometimes private possession is restricted to the house or chalet, the right to the ground upon which it is built lapsing with the existence of the house. In order to limit the number and durability of these buildings, many restrictions are imposed, such as the prohibition to build houses of stones, or chalets of wood cut from trees not belonging to the builder himself or to the corporation to which he belongs, etc.

[1] Miaskowski, *Die schweizerische Allmend*, pp. 18 and following.

7. *Summary*.—This long series of modifications, the result of which was the transformation of primitive communities of goods into the modern forms of public and private property, was accompanied throughout by degenerative changes. The establishment of family property entailed the curtailment of tribal and clan rights. Family property passed into property held by the village; next the development of feudal tenure involved the degeneration of the old agrarian communities; finally, the primitive organization of property with the administrative and political institution dependent on it, disintegrated and disappeared as the primitive community of goods lapsed into the personal enjoyment of these by individuals, and as the primitive method of land tenure passed into the rights of private property.

We see then that degeneration has always accompanied evolution: the destruction of old institutions is involved in the formation of new institutions.

PART II

DEGENERATION IN THE EVOLUTION OF ORGANISMS AND SOCIETIES

WE have seen that modification of organs and of institutions is always associated with partial degeneration. We have now to show that, similarly, when organisms and societies become modified, degeneration is shown in some of their organs or institutions. This shows again the universality of degenerative evolution.

CHAPTER I

ALL ORGANISMS EXHIBIT RUDIMENTARY ORGANS

ALL existing organisms have lost some organs in the course of their phylogenetic development.

This may be proved in two ways: either there are remaining vestiges of these organs, or else they are to be found in other creatures which may be regarded as ancestors.

1. Rudimentary organs, signs of a degenerative transformation in the organism itself, are either organs which have ceased to be functional, or which have so diminished in importance that their total disappearance would be unattended by any appreciable loss to the organism. In the majority of cases this cessation of function is attended by a corresponding structural decay.
2. The system of comparing living organisms with their presumptive ancestors equally demonstrates the retrogression of certain organs.

Among the Orobanchaceæ for instance, parasitic plants derived from normal green plants, no trace of cotyledons is observable from the period of germination.[1]

Among animals, taking the horse as an example, several organs have wholly disappeared. In the genealogy of the horse, which is well known, the earliest ancestor *Eohippus*, possessed five functional fingers on the fore-feet, and four toes on the hind-feet. The horse still possesses one functional finger and one functional toe, two rudimentary fingers and two rudimentary toes. Two fingers and two toes have entirely disappeared.

It is hardly necessary to point out that this system of comparison does not demonstrate the

[1] L. Koch, *Die Entwickelungsgeschichte der Orobanchen.* Heidelberg, C. Winter, 1887.

degenerative changes attending the phylogenetic development of an organism with such incontestable certainty as does the existence of vestiges of rudimentary organs.

It is our belief that all organisms contain vestiges of organs, either more or less apparent. In our present condition of knowledge, however, it is quite impossible — particularly as regards plants — to prove this theory universally. It is to be hoped, however, that future researches will ultimately succeed in establishing it.

In the meantime, we will point out the most typical among the cases known to us.

With animals, as with plants, our investigations have extended not only to every kind of group, but to the most varied systems of organs, thus giving our theory an extremely wide application.

SECTION I.

Rudimentary Organs of Animals.

§ 1. *Rudimentary Organs in Man.*

Throughout the whole human organic systems signs of degeneration abound.

1. *The Integumentary System.*—In the ancestors of man, the entire surface of the skin was covered with hairs. Man's clothing of hair is far from perfect, the hairs of which it is composed being rudimentary.

According to Hertwig, the teeth should be regarded as part of the tegumentary system, as they really represent the scales of the skate, situated within the buccal cavity.

In man, the last molar, or wisdom tooth, is a rudimentary tooth. The small-sized shallow crown, the diminished number of tubercles, the fusion of the roots, the tardy appearance and occasional absence altogether, are all indications of a rudimentary condition.

2. *The Skeleton.*—With few exceptions, the articular surfaces of the bodies of mammalian vertebrates are covered in youth with bony plates. These sometimes become very thick, and are called terminal epiphyses. In some mammals—the Sirenians, for instance—the terminal epiphyses have disappeared. In man they still exist, but in an advanced stage of degeneration. In the lower vertebrates, such as the crocodile, many more ribs are functional than in man. In the crocodile all the ribs connected with the cervical vertebrae are functional, whereas in man they have degenerated. Of one entire section of the human vertebral column—the tail—so fully developed in the majority of other vertebrates, only a vestige now remains.

Other rudimentary skeletal pieces are the lesser horn of the hyoid bone, the stylo-hyoidean ligament, the coracoid process, and the interclavicular ligament.

3. *The Muscular System.*—The cutaneous muscles, those of the shell of the ear, and those that move the tail, which in most mammals are well developed, are still present in man, but have degenerated.

Further, there is to be found in man the intra-acetabular part (the round ligament) of the deep flexor muscle of the toes which is functional in some animals—in young ostriches, for instance. In the adult ostrich the intra-acetabular part is separated from the rest of the muscle, which is attached to the pelvis. Traces remain in the horse of a connection between the intra- and the extra-acetabular parts; the muscle itself is divided into two parts, the pectineal muscle in the thigh, and the deep flexor muscle of the toe situated in the leg. In the orang-outang, this degeneration has made further advances than in man, the intra-acetabular part of the muscle having entirely disappeared.[1]

4. *The Nervous System.*—Here we find numerous signs of degeneration, of which the following are a few examples:

In the brain the pineal gland, the last remaining vestige of what was formerly a functional eye, is present.

In the spinal cord the *filum terminale* still exists. We know that the spinal cord in man does not retain its normal thickness to the extremity of

---

[1] See Sutton, *Ligaments, their nature and morphology.* London, 1887.

the vertebral column but is arrested at the first lumbar vertebra. There a considerable number of special nerves leave it, forming a mass of branches like a horse's tail. Along the centre of these nerves, in the middle line, a slender filament represents the spinal cord to the extremity of the coccyx. This filament is the *filum terminale*, the spinal cord in a condition of degeneration.

5. *The Digestive System.*—The cæcum and its vermiform appendage, are well known to be organs which have degenerated.

6. *The Vascular System.*—In quadrupeds the intercostal veins are vertical, the blood consequently flowing against gravity. These veins contain valves which indirectly facilitate the upward and onward flow of the blood by preventing it from running back. Man, being a biped with a vertical thorax, is provided with intercostal veins that are almost horizontal. The ancestral valves being no longer indispensable are in a condition of degeneration.

7. *Sense Organs.*—In the olfactory organ there remains a degenerate *Jacobson's* organ. In the organ of sight there is a third eyelid in a state of degeneration. In the organ of hearing there remains on the shell of the ear a kind of point (Darwin's point) which is the last remaining vestige of the ancestral elongated and pointed ear.

8. *Genito-urinary System.*—There is a whole series of rudimentary organs in the genito-urinary system of the higher animals. As is well known,

the Wolffian body plays a considerable part in the formation of the system. This body, the primitive kidney, loses its urinary function at a certain stage of embryonic development, and the permanent kidney which gradually develops alongside, assumes the urinary function. Later on, the Wolffian body assumes new functions, becoming an important part of the genital apparatus.

In this transformation partial degeneration occurs, resulting in such reduced structures as the epididymis, the organ of Rosenmüller, the *vas aberrans*, etc. (see fig. 57).

§ 2. *Rudimentary organs in various groups.*

1. *Cœlenterates.*—The Cœlenterates comprise three great groups.[1] The *Anthozoa*, of which the coral is a type, the *Hydrazoa*, which include fresh water Hydra and the common jelly-fish of our seas, and the *Ctenophora*, of which the chief representative in our seas is Cydippe, a globular transparent animal frequently to be found floating in large numbers on the surface of the water.

The colonies of *Anthozoa* are usually composed of individuals all exactly alike. In some species, however, in the Pennatulidæ and the Alcyonaria for instance, there is a distinct differentiation amongst the numerous individuals composing the colony. Side by side with sexual individuals

[1] See C. Vogt and Emile Yung, *Traité d'anatomie comparée*, vol. i.

provided with tentacles and the eight mesenteric folds, are other far simpler individuals: the zooïdes, the function of which is respiratory and of which the greater part of the organs have degenerated; the generative organs are lacking, the tentacles are very small, and the mesenteric folds only number two instead of eight. Degeneration, then, is exhibited side by side with specialization.

Among the *Hydromedusæ* similar examples abound. It is known that the polyp-like or medusa-like forms of this group which may live independently, frequently associate themselves together to form colonies, sometimes predominantly polypoïd, sometimes completely medusoïd, and occasionally a mixture of the two.

In these cases a marked polymorphism is often apparent. The different individuals become adapted to definite functions, and the corresponding organs undergo special development; the other parts of the body having become either unnecessary or merely accessory, begin to degenerate and finally disappear. Thus we see in Hydroïd colonies, not only the hydra-like members, nutritive, fixed and sterile, and the medusa-like members which are reproductive and become free from the colony, but also certain individuals which are termed gonophores. These gonophores are really medusa-like members which have lost their independent movement, and have consequently more or less lost both their tentacles and their umbrella-like discs, *i.e.* their organs of locomotion. In some colonies of Hydroïds, polymorphism has made such advances that there are tactile individuals of which the digestive tube lacks both mouth and tentacles, and other purely defensive individuals of which the internal organs are almost all in a state of atrophy.

Opinions differ regarding the complicated question of the structure of the Siphonophora. The organism (fig. 52) may be regarded as a simple medusa of which the different appendages— the pneumatophore (*pn.*), the swimming bells (*cl.*), the siphons (*s.*), the shield (*b.*), the tentacle (*t.*), the palp (*pa.*), the gonophores (*go.*), etc.—constitute the organs, or as a colony each part of which is represented by an individual polypoïd adapted to fulfil a special function. Whatever theory is accepted, it is clear that a whole series of parts of the creature must be regarded as organs in a condition of degeneration.

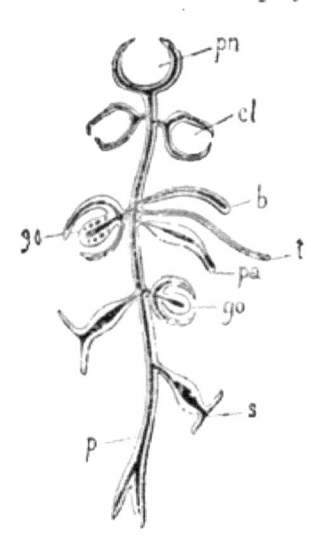

Fig. 52.—Diagram of the structure of one of the *Siphonophora*. *pn*, pneumatophore or float; *cl*, swimming bell; *b*, protective polyp; *t*, tentacle; *pa*, palp; *g*, gonophore; *p*, peduncle bearing the mouth; *s*, individuals for nutrition.

We accept the second of these theories,[1] and

[1] Haeckel, *System der Medusen*: Jena, 1880-1881. A. Lang, *Traité d'anatomie comparée.*

regard the whole creature as formed from a Craspedote Medusa which has become mother of a colony and of which the umbrella, developed into the pneumatophore ($pn.$) has had its radial canals greatly simplified and its tentacles reduced to one during the growth of the colony. The stalk-like stomach ($p.$) of this medusa has increased in length, but this development is attended by corresponding degeneration, the buccal aperture, which is situated at the free end of the peduncle in normal Craspedote Medusæ, having entirely disappeared. The stalk, formed in this way, serves as a support to the number of other individuals of which the colony is constituted, and which are remarkable for the great morphological variation they exhibit. Among these individuals, those at the top, *i.e.* those nearest to the air-sac, fulfil the function of locomotion. They become transformed into swimming bells ($cl.$) and contain no organs whatever. Below these locomotory organs are the sexual individuals, or gonophores ($g.$), and the sterile individuals ($s.$). The former are of medusoïd structure, the umbrella is more or less perfect, and they are sometimes provided with tentacles, and possess a peduncle or manubrium which sometimes has a buccal aperture. The sterile individuals provide nutrition for the others. The organs no longer essential to them atrophy in a variable degree. In the case of some the umbrella is present, in others it is absent, and between these two extreme types come intermediate types which exhibit every possible stage of degeneration.

In the Ctenophore group development and degeneration are exhibited simultaneously in the organs of locomotion. The fundamental and typical shape of the Ctenophore is round or oval, and the eight sides are provided with swimming plates, originally uniform—as in *Beroë*. The individuals belonging to this group exhibit important evidences of modification in their external morphology. The body being sometimes compressed in various directions, the shape is altered from the original, and assumes a more or less irregular appearance. The organs of locomotion undergo a corresponding change. Take for example an adult specimen of *Bolina norvegica* (fig. 53); the body is lobate, although it was round during the larval period; the swimming plates are not uniform, four being long and reaching along the whole length of the body, the other four being only developed in the upper half of the body as far as where the lobes are inserted, where they end as degenerate hair-like processes. By referring to the Cestidæ, which are ribbon-like in shape, it will be seen that by means of compression the body

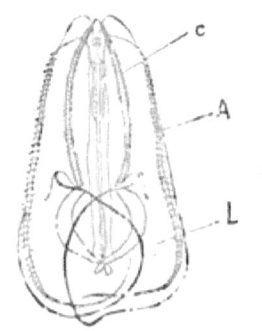

FIG. 53.—*Bolina Norvegica*, seen from the broad side. C, short rows of swimming plates; A, long rows; L, lobes. (After Vogt and Yung, *Traité d'anatomie comparée.*)

of the animal is lengthened out into the shape of a narrow ribbon. Of the primitive lateral row of plates, four are represented by mere vestiges, and the other four, which continue to be functional, are situated, closely coupled together, on the two edges of the animal.

2. *Worms.*—We will next take in succession the Plathelminthes or flat worms, the Rotifers, the Nemathelminthes, or round worms, the Annelids and the Gephyreans.

Among the *Plathelminthes*, the group of *Cestodes* contains the common tape-worm of man (*Tænia solium*). In the course of its parasitic existence this worm has undergone considerable morphological changes. The digestive tube is lacking, and the whole nervous system has become greatly simplified. The degeneration of the nervous system is not, however, complete, for important vestiges still persist. The degeneration of the digestive tube is much more thorough. In *Tænia solium* it is altogether absent. In species closely allied to the *Tænia* some slight vestiges of the digestive apparatus yet remain. The head or scolex of some species of *Tetrarhynchus* contain glandular cells which have been homologized with the salivary glands of other flat worms (*Trematodes*). In other species of *Tetrarhynchus* there is a rudimentary organ which represents the oval sucker of the *Trematodes*, and in *Anthrocephalus elongatus* the orifices of the salivary glands are in the region of this vestige of the digestive tube.

Many facts in support of our argument may be drawn from a study of the excretory organs of the Cestodes.

The condition of these organs in *Caryophylleus mutabilis* best represents the primitive condition. Within the body of this worm are a large number of narrow ducts with ciliated funnels communicating with the spaces in the parenchyma (Fraipont). These organs communicate with canals which gradually reunite and anastomose to find a vent in one single aperture, the *foramen caudale*, which is situated in the posterior part of the body where there is a bladder. In the Cestodes, however, where the body is very long, the action of the bladder is insufficient to secure a complete evacuation. Secondary apertures are therefore formed at intervals along the main ducts. This new structure entails the degeneration of the terminal bladder which has become superfluous. In *Botryocephalus punctatus*, which possesses a great number of excretory apertures, the primitive evacuatory apparatus—*i.e.* the contractile cavity—has completely disappeared.

Rotifers are minute animals, usually living in fresh water, a few being marine. One of them, an inhabitant of damp earth or moss, has been supposed to possess the power of revivifying after complete dessication. At the anterior end of the body, a rotifer possesses a complicated ciliary apparatus which fulfils the function of locomotion, and from the rotatory movements of which the name of the group have been given. This organ is well developed in those types which lead an independent existence, but in those where movement is more restricted or where the character of the organ has changed, it is considerably modified and reduced in size. In *Philodinæ* (crawling Rotifers) the organ of rotation has lost the central part, and in its place is substituted a very complicated organ of prehension.

In sessile forms such as *Floscularia*, the organ is modified and only the primitive character of the inner ring is retained, while the outer ring is segmented and becomes a series of arms or lobes, furnished with stiff bristles. In *Apsilus*, another sessile form, the organ of locomotion has disappeared; this is obviously an instance of true retrogression, for in young specimens a vibratory crown still persists.

The *Nemathelminthes* contain such round worms as these common intestinal parasites: *Ascaris, Oxyuris, Strongylus*, etc. These all belong to the Nematode group, and possess a complete digestive tube. *Gordius*, however, exhibits organs which are reduced in a marked degree; in the adult animal the buccal orifice of the digestive tube is closed, and the posterior part of the intestine has disappeared, although in the young worm the alimentary canal is complete. This modification, though incomplete and appearing only in the adult life of *Gordius*, is complete and permanent in some other Nematodes. In *Echinorhynchus*, for instance, the digestive tube is absent, and nourishment is obtained by means of osmotic soaking through the body walls.

The *Annelids* comprise the annulated sea-worms and forms like the common earth-worm (*Lumbricus terricola*). In these creatures we will take, first, the development of the eyes. In *Oligochætes*, which for the most part live in soil or mud, the organs of sight are greatly reduced. The *Naïdomorphæ* alone have eyes. The *Archiannelida*—*Histrior* (a parasite), for instance—possess eyes when young, but in the adult state the eyes have greatly degenerated. As a rule, the eyes of the *Polychætes* are well developed, and in some of them quite remarkably so. In species, however, which do not move about much, the eyes are merely represented by small pigmented spots.

We may mention, too, the *Gephyreans*, without pledging ourselves as to their exact relationship. *Bonellia viridis*, the history of which is well known and of great interest, belongs to this group. The male Bonellia lives as a parasite on the proboscis, or in the gullet or the nephridium of the female. It is flat and small, and has neither mouth, arms, nor circulatory system. All the organs remain as in the larval condition, with the exception of the genital organs, which are fully developed.

Bonellia and *Dinophilus*, a rotifer, of which the male is degenerate, exhibit a progressive degeneration of all the organs not connected with reproduction. Degeneration has made furthest advances in *Bonellia*, which affords a striking example of **retrogression** side by side with development.

Investigations of animal series such as these might well be continued throughout the various classes and groups, showing the existence of rudimentary organs in all. We will restrict ourselves here, however, to mentioning the larger subdivisions only, taking one example from each group.

3. *The Echinoderms.* — This order comprises star-fish or asterids, sea-urchins, Crinoïds and Holothuria or trepangs.

Of these we will take the star-fish, and proceed to examine its digestive tube. Under normal circumstances, the intestine terminates in a dorsal anus, centrally, or slightly excentrically placed, preceded by a very short but well-developed rectum. *Asteracanthion* and *Solaster* furnish good examples of this. In some kinds of Asterids—in *Astropecten aurantiacus*, for instance—the anus no longer exists, and the rectum, having become useless, is greatly reduced, though still exhibiting signs of its original condition.

4. *Mollusca.* — Of the group of Mollusca we will take the Gastropods and the Cephalopods as examples.

There are two kinds of Gastropods, straight and twisted, the former representing the primitive type. The straight types—such as *Chiton, Patella*,

Haliotes, and *Fissurella*—are bilaterally symmetrical, while in other Gastropods the spiral twisting of the body causes a progressive diminution of the organs situated on the side towards which the twisting occurs; the organs of the left side may therefore become smaller, and finally atrophy almost completely.

The internal shell of the Cephalopods furnishes a striking example of a rudimentary organ.[1] The Nautilus (fig. 54) has a shell, the spiral coils of which are pressed tightly against one another. The spiral is divided into a series of chambers by means of partitions, each partition being provided with an aperture for the admission of the siphon (*l.*). The shell of *Spirula* (fig. 55 and 56, A), a creature still existing, is only partially curled round; the last chamber of the shell is very small, and only encloses a part of the animal, most of which remains outside the shell, and partially covers it by the mantle (*p.*), the shell being therefore partly external and partly internal. On examining the fossil species : *Spiru-*

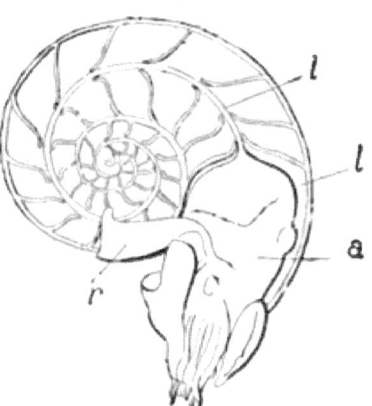

FIG. 54.—*Nautilus pompilius.*
l, terminal chamber of the shell ; *a*, body of the animal ; *l*, siphon ; *r*, mantle fold. (After Owen.)

---

[1] W. Boas, *Lehrbuch der Zoologie*. Jena, 1894.

lirostra (fig. 56, B), *Belemnites* (fig. 56, C), and *Conoteuthis* (fig. 56, D), a progressive simplification of the shell may be observed, the latter becoming less and less coiled, until finally the original shell is transformed into a straight chambered portion surmounted by a stiliform process. Degeneration is principally exhibited in

FIG. 55.—*Spirula prototypos.*
a, body of the animal; *p*, mantle fold; *c*, shell, partly internal, partly external. (After Owen.)

FIG. 56 —Shells of various Cephalopods.
A, *Spirula*; B, *Spirulirostra*; C, *Belemnites*; D, *Conoteuthis*; E, *Ommatostrephes*; F, *Loligopsis*. (After Boas.)

the segmented part, which becomes more and more reduced. In *Ommatostrephes* (fig. 56, E), and in all existing species (such as *Loligopsis*, fig. 56, F), with the exception of those just mentioned above, degeneration has become complete. The shell no longer contains a cavity for the reception of the animal, and the phenomenon already mentioned with regard to *Spirula*—the development of the organism round about the shell—has become more marked in character. The shell has become internal instead of external, and forms the so-called cuttle-bone. This structure, being only the vestige of what was originally an external shell, must be regarded as a reduced organ.

5. *Arthropoda.*—The group of Arthropods comprises the Myriopoda, the Crustacea, the Arachnida, and the Insecta.

Instances of rudimentary organs are very common among the Crustacea, but our investigations with regard to the appendages of the cray-fish were so thorough that we will give examples from another group, that of the Insecta, instead.

Insects are characterized by the possession of three pairs of legs and two pairs of wings. The organs of flight exhibit a multitude of special adaptations, and numerous instances of degeneration are exhibited. In the *Neuroptera* (dragon flies), *Hymenoptera* (saw flies and bees), and *Lepidoptera* (moths and butterflies), the four wings are generally all alike and fully developed. In the *Diptera* (flies), the posterior wings have atrophied and are represented by two reduced organs, the "balancers," these being absent in certain types. The *Strepsiptera* form a group of which comparatively little is yet known; the larvae live in the nests of bees; the females have no wings; in the males the anterior wings are rudimentary and the posterior wings are fully developed. In the group of *Coleoptera* (scarabs, cockchafers, longicorns, and beetles), the anterior wings ("wing covers") have become very resistant, and constitute a perfect cuirass covering and protecting the abdomen, and the posterior wings, the organs of flight, are folded beneath the wing covers when in repose. In the Scarabæi and the Calosoma the posterior wings are much reduced. In other Coleoptera the wing covers are united by their inner edges, thus rendering all movement of the underlying wings ineffectual and useless. The result may, as in *Gibbium*, be the total atrophy of these under wings. In the Staphylinidæ, the upper wall of the abdomen is so strong that the protection of wing covers is unnecessary; consequently the wing covers have degenerated into little lamella which cover only the anterior quarter of the body. In the female *Lampyris* (glow-worm), the wings have totally disappeared. The *Orthoptera* (cockroaches and earwigs) exhibit a great number of variations in the organs of flight. In cockroaches all four wings are usually fully developed; in some specimens, however, and these principally female, a very pronounced degeneration of all the wings may be observed. The anterior wings of the Forficulides are reduced. In the group of Phasmidæ we find side by side with species of the genus *Bacillus*, which have no wings and look like dried twigs, *Phyllium siccifolium* to which the large green and yellow wings give the appearance of a leaf. The neuters of Termites have no wings.

6. *Vertebrates.*—Among Vertebrates the genito-urinary system contains a great number of rudimentary organs. In order to fully understand the nature of these organs, it will be necessary to glance through the ontogenetic and phylogenetic development of the system.

The first stage in the formation of the kidney-system is the *pronephros* (fig. 57, A). This primitive organ consists of intricate canals (a, a, a), opening into the body cavity at the point where the glomeruli are formed on the sub-intestinal vein. All these canals originally had apertures to the exterior. Later on, however, these uriniferous tubules became connected with one single excretory canal (c.) opening into the cloaca (ce.). The primitive genital gland was situated close to the pronephros. In process of time the mesonephros replaced the pronephros (fig. 57, B); in origin it was quite distinct from the pronephros, its appearance being that of a secretory urinary gland (G.) and its secretory canal (c.) (segmental duct), was the same as that of the pronephros. The urinary system thus formed, became and still continues to be, closely connected with the genital gland, the discharging canals of which passed through the mesonephric kidney in order to find a passage to the exterior through the segmental duct. During the mesonephric stage, another canal was formed which started from the cloaca and opened out into the general body cavity, this was Müller's duct (M.).

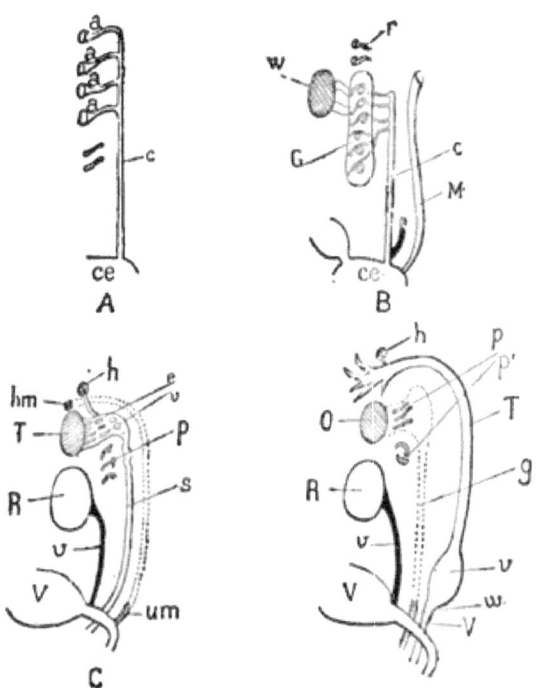

Fig. 57.—Development of the Urino-genital system in higher Vertebrates.
A. Pronephric stage. a, tubules; c, excreting duct; ce, cloaca.
B. Mesonephric stage. G, mesonephros; r, remains of pronephros; c, excreting canal; w, neutral genital gland; M, Müller's duct; ce, cloaca.
C. Metanephric or adult stage, in the male. R, permanent kidney; U, ureter; V, bladder; T, testis; e and r, epididymis and vas deferens; s, vas deferens; hm, hydatid of Morgagni; h, hydatid; p, paradidymis; um, uterus masculinus.
D. Metanephric or adult stage in the female. R, permanent kidney; U, ureter; V, bladder; O, ovary; p and p', parovarium and paraophoron; w, Weber's organ; V, vagina; u, uterus; t, aperture of Fallopian tube; h, hydatid.

The mesonephros, however, was not the permanent kidney. In the course of time the *metanephros*, the permanent excretory gland, was developed (fig.

57, C and D). This development was attended by important modifications, further instances of degeneration taking place, and fresh organic connections being established. An examination of the male and female sexes is necessary in order to explain what really took place.

In the male (fig. 57, C) the mesonephros began to atrophy; that part which was connected with the testes was transformed into the epididymis and the *vas deferens* (e. and v.) of the true male genital organs; the remaining part atrophied, and when the permanent organization was attained, only persisted in the form of a paradidymis (p.) and a hydatid (h.), organs which are quite without function in the adult state.

The discharging canal which, during the mesonephric stage was common to both urinary and genital glands, remained simply in connection with the testes, and then became the *vas deferens* (s.) of which the cloaca having disappeared, the terminal extremity became gradually individualized. The permanent kidney (R.) became connected with a fresh canal—the ureter (u.) which was formed by degrees at the expense of the primitive discharging canal, and subsequently became separated from the latter in order to empty itself into the bladder (v.).

These changes were attended by a remarkable evolution of Müller's canal, which first increased in size, and then at a certain point proceeded to atrophy until all that remained were the distal and proximal extremities in the shape of reduced organs, the hydatid of Morgagni (*hm*), and the uterus masculinus (*um*), neither of these being functional. The intervening part of the canal remained, and formed a canal which has been described by Gasser.

We see then, that in the genito-urinary apparatus of an adult male there are:

(1) Organs which have come into existence at different times, but which have retained their original functions, viz.: the testes, the kidney (metanephros), and the ureter; (2) Organs which are functional, but of which the ultimate function differs from the original, viz.: the epididymis and the *vas deferens*; (3) Reduced organs, vestiges of what were formerly active organs, viz.: the hydatid and the paradidymis; (4) Reduced organs, vestiges of Müllerian canal which only became active in the female, viz.: the hydatid of Morgagni and the male uterus.

In the female (fig. 57, D), the development of the renal part is similar to that we have just described in the male. Taking first that part of the mesonephros which became connected with the genital gland and the corresponding discharging canal, we find that the canal disappeared with but rare exceptions in which it formed Gärtner's duct (*g.*), the lower part persisting in the form of a rudiment (Weber's organ (*w.*)); the upper part became reduced to a small tissue which surrounded the paraovarium (*p.*), and the paraophoron (*p.*) vestiges of what was formerly the mesonephros. Müller's canal became considerably enlarged; it formed the vagina, the uterus, and the Fallopian tubes; at the upper end it was connected with the hydatid, a vestige of the mesonephros.

It is plain then that the genital-urinary apparatus of the female comprises some organs of which the functions remain unchanged: the ovaries, the permanent kidney, the Fallopian tubes, the uterus, the vagina, and the ureter, and some rudimentary organs, vestiges of what were once active organs; the paraovarium, the paraophoron, hydatid and Weber's organ.

The complicated development of this system becomes clear if a careful study is made of the history of the genito-urinary apparatus of the entire series of vertebrates.

It appears that the various phases through which the embryos of the higher vertebrates pass are stages similar to those which may be observed in the adult lower vertebrates.

The principle of recapitulation, that the embryonic stages of higher animals recapitulate successive stages attained by the adults of lower animals, receives a full corroboration from the facts we have been displaying.

Amphioxus, for instance, remains still at the pronephric stage: fish as a rule have a mesonephric or permanent kidney. Some lizards (*Lacerta*) up to the age of two years make use of the mesonephros as the organ for eliminating urine, but, at the same time they make use of the metanephros which is also functional.

In *Chamæles* the mesonephros remains partially active throughout life. Both birds and mammals completely lose the mesonephros, and in the adult stage the metanephros is the only active kidney.

This is not the place in which to complete our study of the recapitulation theory, and we shall have to recur to it later on; but it was impossible to describe the numerous rudimentary organs of this system without taking a comprehensive glance at the individual and specific development of the whole. This investigation, moreover, raises another question.

It has just been shown that the epididymis is only a vestige of the mesonephros, but in this case it cannot be said that there has been degeneration; what has happened is that an organ has been transformed, and that one function has been replaced by another. According to some authorities the suprarenal capsule, an organ of unknown, but doubtless essential function, is the result of the transformation of the pronephros. If this theory be ultimately established, it will furnish a second example of what we have stated above. The thyroid gland may and ought to be investigated from this standpoint. The various component parts of this organ had no original connection. In higher vertebrates only the central part of the organ appears to be similar to the same organ in the whole series of vertebrates. In man the mesial part of the thyroid gland is reduced, but it cannot correctly be said that the organ itself is degenerating. The contrary may even be asserted, for all we know upon the subject goes to prove that in mammals the thyroid gland is formed and established at the expense of the primitive rudimentary organ of which all the morphological and embryological connections are changed. This secondary development, entailing the loss of a reduced ancestral organ, is attended by a functional modification of great importance. The primitive function probably discharged by the gland has given place to another and rather vague function, but one which is connected with the breaking down of toxic matter formed by living tissues.

An examination of the vertebrae of vertebrates will show the existence of rudimentary organs throughout the whole group, each type and each individual among the vertebrates exhibiting special degeneration. In order to briefly demonstrate this point, take two quite different types of which all the vertebrae are well known and can therefore be examined without difficulty: man and the frog.

We know that the construction of the primitive bony vertebra was as follows:—a centrum, carrying neurapophyses, an intercentrum, carrying hæmapophyses, and a pair of unforked ribs.

Each phase in the evolution of this primitive vertebra has been attended by degeneration.

In man the vertebral column consists of:—

    (*a*) Seven cervical vertebrae.
    (*b*) Three dorsal vertebrae.
    (*c*) Five lumbar vertebrae.
    (*d*) Three sacral vertebrae.
    (*e*) Six coccygeal vertebrae.

Each one of these vertebrae exhibits important modifications, and shows signs of degeneration. The proatlas is represented by its intercentrum only. The atlas consists of a centrum and neurapophyses, but there is no zygapophyses; it possesses one pair of ribs and hæmapophyses in a reduced condition. The axis consists of the same elements but carries postzygapophyses.

The four following cervical vertebrae consist of the same elements, but carry both zygapophyses and postzygapophyses. In man all the cervical vertebrae, with the exception of the first (the proatlas) have lost the intercentrum. The next vertebrae which is generally regarded as the seventh cervical vertebra, consists of the same elements, but it ought to be regarded as the first dorsal vertebra. The vertebral artery and the sympathetic nerve trunks accompanying it, do not pass through the vertebral canals; in some cases these do not exist. The so-called seventh cervical vertebra has occasionally one fully developed rib which articulates with the sternum like the ribs of true dorsal vertebrae.

There are thirteen dorsal vertebrae. With the exception of that which has just been described, they consist of:—a centrum, neurapophyses, and a pair of fully developed ribs; the intercentra has completely disappeared, and the hæmapophyses, which form the head and neck of the rib, are in a reduced condition. In the five lumbar vertebrae which follow, the rib disappears, or to speak more accurately, the transverse processes are all that remain of what were the ribs, and have ossified with the vertebra.

The sacrum is a region profoundly modified to support the pelvic basin. It is formed by the fusion of five vertebrae, each consisting of a centrum, neurapophyses, and short bicipital ribs. The first three vertebrae are the true sacrals, as these alone support the basin; the two following are really caudal vertebrae in process of fusion with the sacrum. In monkeys other than anthropoids there are really only three sacral vertebrae and these are at once succeeded by the tail. In man, the tail[1] consists of six vertebrae of which the two first—which consist of a centrum and neurapophyses and of bicipital ribs—have fused with the sacrum, while the four remaining lower vertebrae, which consist of only the centra (the first still exhibits rudimentary neurapophyses) have fused, and form the coccyx.

[1] See Albrecht, *La queue chez l'homme* (Bull. soc. Anthrop. Brux., vol. iii., 1884-1885.

2. *The Frog.*—The vertebral column in the frog consists of nine vertebrae and the coccyx (urostyle).

The first vertebra (the proatlas), which is fully functional, retains a centrum and two well-developed neurapophyses, but the transverse processes, the intercentrum, the hæmapophyses and the ribs have all disappeared.

The eight following vertebrae each have a centrum, neurapophyses and transverse processes which at least partially represent the ribs.

The coccyx, whether it be formed by the lengthening out of the last caudal vertebra or by the fusion of several, is undoubtedly part of the vertebral column which has been transformed. The coccyx of the frog is equal in length to the whole of the remaining part of the vertebral column, and is fully functional; it serves as a support to the pelvic region and fulfils the part of a sacrum from the physiological point of view. The coccyx, excepting at its commencement, consists of only one centrum, or of several fused centra, all other elements of the vertebrae have disappeared. Here then is an animal in which the modifications of the vertebral column have been attended by the following retrogressive phenomena: in the upper part of the vertebral column from three to five parts of the vertebrae have disappeared; in the lower half, all the parts, excepting one, are gone.

Section II.

Rudimentary Organs in Plants.

We have just glanced through a series of rudimentary organs in animals, and many more examples might easily have been furnished, but, when dealing with rudimentary organs in the vegetable world, much greater difficulty is met with. In plants, the elimination of non-functional organs is usually complete, and the vestiges left are insignificant and hard to recognize. We can find, however, amongst the various groups of the vegetable world, and especially among the Phanerogams, some instances of reduced organs.

§ 3. *Rudimentary organs in various groups of plants.*

1. *Algae.*—On the surface of sea-wrack (*Fucus*) may be found, distributed in large numbers, little crypts (conceptacles) with hairs growing out of them. On certain parts of the plant, these crypts represent the organs of reproduction, producing eggs and spermatozoa; in other parts they fulfil no known function and may be regarded as conceptacles arrested in the course of development. The fact that in other Fucaciæ (*Splachnidium*), fertile conceptacles are distributed over the entire surface of the plant adds support to this theory.

2. *Mushrooms.*—Among the Peronospora and the Saprolegneæ there originally existed, besides the various sexual means of propagation, a typical reproductive process, including eggs, and antheridia, consisting of male branches separated by a cell-wall from the rest of the organism.

In *Pythium,* for instance, in which this primitive stage may be observed, an actual fecundation takes place, the protoplasmic contents of the antheridia being injected into the ova.

In other species, the organs of reproduction have undergone a more or less complete degeneration. In *Phytophthora,* a small part only of the male protoplasm passes into the ova. In some species of *Saprolegnia* and *Achlya,* the male branch continues to attach itself to the ova, but the membrane between them remains intact, and consequently protoplasmic communication is not established.

In other species, the antheridia are very short, and do not even touch the female cells.

In *Leptomitus,* which exhibits an advanced stage of atrophy, the female organs are not discernible, and reproduction is carried on completely asexually.[1]

3. *Bryophyta.*—In the germination of a certain number of Hepaticæ, belonging to such widely

---

[1] For further details see A. de Bary, *Vergleichende Morphologie und Biologie der Pilze*, Leipzig, Engelman, 1884. W. Zopf, *Die Pilze*. In Schenk's *Handbuch der Botanik*, 4. Bd., Breslau, Trewendt, 1890.

separated genera as *Blasia*, *Radula*, and *Preissia*, four cells of equal size are formed, arranged round a centre. One only of these cells proceeds to develop into a plant, and the others simply atrophy. In all probability the **Hepaticæ** have sprung from some ancestor, in which each spore gave rise to four individuals.

4. *Pteridophyta.*—According to Farlow,[1] instances of apogamy—the loss of sex—such as have just been described as existing among mushrooms, are also exhibited in certain ferns. In some species,[2] the eggs are not fertilized, but the organs of reproduction still persist in a reduced condition; in other species there are no spores, and the prothalli spring directly from the leaves (apospory).

5. *Phanerogams.*—Some of the Phanerogams— *Silene* (fig. 58), *Melandryum*, (fig. 59), *Asparagus*, etc.—exhibit unisexual flowers, but have obviously sprung from species of which the flowers were hermaphrodite.

In *Silene maritima* (fig. 58) there are hermaphrodite flowers (fig. 58, B), and also unisexual flowers. The female flowers (fig. 58, A) still possess some tiny stamens, each of which is provided with filaments and anthers in a state of degeneration. The male flowers have non-functional pistils,

---

[1] W. Farlow, *Ueber ungeschlechtliche Erzeugung von Keimpflänzchen an Farnprothallien*. Bot. Zeit., 1874, p. 180.

[2] A. de Bary, *Ueber apogame Farne u.s.w.* Bot. Zeit., 1878, p. 449.

consisting of an ovary stylus and stigma in reduced conditions.

In *Melandryum* (fig. 59, A) the female flowers retain only vestiges of stamens, and the pistils of the male flowers are reduced to mere filaments. In *Asparagus officinalis* all the transition stages between herma-

Fig. 58.—Flowers of *Silene maritima*. A, female flower with rudimentary stamens; B, hermaphrodite flower.

phrodite and unisexual flowers may be observed: in the unisexual flowers, the organs of the opposite sex still exist, though in various stages of degeneration. In many unisexual flowers which have sprung from hermaphrodite flowers — *Valeriana dioica*, for instance — no traces of the non-functional organs remain.

Fig. 59.—Flowers of *Melandryum album*. A (to the left), a female flower; to the right, a male flower; e, rudimentary stamens forming a circle at the base of the ovary.

Among the Phanerogams, rudimentary organs

appear not only in the reproductive organs, but in the accessory organs of the flower—the calix and corolla. Many of the Umbelliferae exhibit a reduced calyx.

The corolla persists, though in a very reduced state in cleistogamous flowers (*i.e.* flowers which never open, and which are self-fertilizing)—such as the *Oxalis*, *Impatiens*, Violet, etc. The corollas of the winter-opening flowers of *Stellaria media* are much reduced, and for a very obvious reason— the corolla exists only for the attraction of insects, and there are no insects at that time of the year.

§ 4. *Reduced organs in the vegetative apparatus of the Phanerogams.*

We have seen that reduced sexual organs are exhibited among the various groups of plants, and we will now mention a few instances of reduced organs in the vegetative apparatus of the Phanerogams.

1. The embryo within the ripe seed of Phanerogams contains a rudimentary root which develops during germination. In certain Nymphaeaceae—*Nelumbium Euryale* and *Victoria*—this root never properly develops.

In other aquatic plants degeneration has gone further; in the embryo of *Utricularia*, for instance, the root has entirely disappeared.[1]

[1] For further details relating to the roots of the Nymphaeaceae and the *Utricularia* see Goebel in *Pflanzenbiologische Schilderungen*, vol. ii., Marburg, Elwert, 1891-1893.

2. As a general rule, the cotyledons, which are the two first leaves to appear after germination, are formed within the embryo. The ripe seeds of some *Anemones*, however, contain no traces of cotyledons.[1]

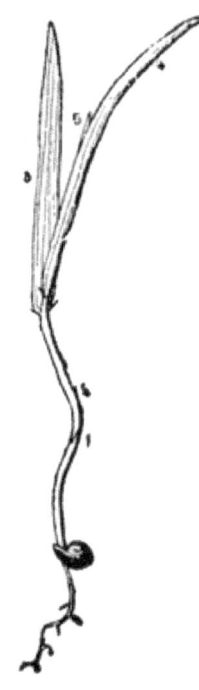

Fig. 60.—Seedling of *Lathyrus Nissolia*.

These are formed, nevertheless, after germination, sometimes sprouting up out of the ground and becoming functional, and occasionally remaining underground, in which case they are quite small, without chlorophyll and nonfunctional (*Anemone nemorosa*); these underground leaves may fairly be regarded as rudimentary organs.

3. In *Lathyrus Nissolia* (fig. 60) there are some very small stipules of unimportant function, at the base of the simplified leaves; these reduced stipules are occasionally absent altogether.

4. The foliage organs in the adult *Oxalis bulpeurifolia* are merely represented by enlarged leaflets. These phyllodes bear reduced leaflets which rapidly disappear. In an adult specimen of the *Acacia* which has phyllodes, these reduced leaves are absent.

[1] E. de Janczewsky, *Études morphologiques sur le genre Anemone.* (*Revue générale de botanique*, t. iv., p. 241.)

5. In several plants the assimilative function of the leaves is lost, either because the plant is parasitic or saprophytic, such as *Corallorhiza, Rafflesia, Cuscuta* and *Orobanche*, or because the assimilative function is relegated to the stem alone as in the *Euphorbia* of the desert, *Ruscus, Mamillaria*,[1] *Phyllocactus, Phyllanthus*, and *Mühlenbeckia*, or to the roots as in *Tæniophyllum*.[2]

In each of these cases the leaves are greatly reduced, and only serve as a means of protection to the functional organs, principally to the flowers and buds, but though very minute they may often be discerned quite easily on the young shoots.[3]

CHAPTER II

SURVIVALS EXIST IN ALL KINDS OF SOCIETIES

It may be said as certainly of societies as of other organisms that certain modifications have taken place, and that no society actually represents a primitive social organization. All have been submitted to more or less important modifications and have lost some of their early institutions in

---

[1] See fig. 51.

[2] See further on the figs. of *Phyllocactus* (fig. 78), *Phyllanthus* (fig. 84), *Mühlenbeckia* (fig. 80), and *Tæniophyllum* (fig. 81).

[3] Goebel, *Pflanzenbiologische Schilderungen*, Bd. i.

process of their development. In many cases this can be historically demonstrated.

It may fairly be asserted that in all societies there are instances of survival, *i.e.* survival of customs, beliefs and institutions, the original character of which has so completely disappeared that they might well be dispensed with altogether.

We shall deal only with such survivals as correspond—*mutatis mutandis*—to the rudimentary organs of animals and plants.

These survivals are of two kinds, the institution itself, such as the various corporations of the city of London, which may still persist, though in a modified condition, or there may remain only traces of the institution in forms, ceremonies, symbols, public games and fêtes, customs and legislative formula.

In order to demonstrate this point, it will not be necessary to make a complete enumeration, furnishing examples from all countries of the world, or to draw up a complete list of survivals in any given country. It will be enough to establish two points :

1. That instances of survival shall be shown to exist in all societies, even where they are least likely to be found.
2. That, in any institution—that of the family, for instance—survivals may be found of all the former stages through which it passed into its present condition.

These two points being established our conclusions drawn from them may be given a very wide application.

§ 1. *Instances of survival in various groups.*

"It is a well-known fact," says Kowalevsky, "that as the past gives place to the present it leaves traces which vary in number and importance."[1]

This is obviously the case with regard to most customs, but it is unnecessary to point out all the instances of survival which abound among the peoples of the countries round about us. They are naturally most common in barbarous societies where the servile imitation of the ancestor plays a much greater part than with us. This is strongly urged by Bagehot in the following passage:—

"Man," he says, "may be defined as a creature of habit. As he has done a thing once, so he will probably do it again, and the oftener he has done a thing the more likely he will be to repeat it in the same way, and, what is more, to insist upon others doing the same.

"By means of counsel and example he transmits to his offspring the customs he himself originated. This is true of the human beings to-day and will doubtless hold good for all time. It is character-

[1] Kowalevsky, *Tableau des origines de la propriété et de la famille*, p. 7.

istic of primitive societies that sooner or later most of these customs come to be regarded as having a supernatural sanction, and the whole community is impressed with the belief that if the old tribal customs are violated, incalculable misfortune will follow."[1]

Social modifications are therefore effected very slowly and with great difficulty—stagnation is the rule, and progress but a rare exception, innovators being forced to retain the greater part of the old institutions, introducing only an inevitable minimum of change. A course of history, or a careful study of the conditions of social institutions at an earlier stage of development than our own, will furnish numerous instances of survivals. It now remains to be seen if there are no rudimentary social groups wherein all the primitive institutions have been retained, and which, having undergone no modifications, exhibit no traces of degeneration. We shall further see if no form of society exists uninfluenced by the spirit of tradition, and where institutions which have come to be regarded as no longer useful, disappear suddenly and entirely either by voluntary dissolution or by legislative measures. Only in these two extreme cases can the existence of survivals be questioned. We have then to establish two points:

(*a*) That all societies, even those to be regarded as

[1] Bagehot, *Lois scientifiques du développement des nations*, p. 154, Bibl. scient. intern., Paris, F. Alcan, 1885.

primitive, have undergone certain modifications.

(b) That all societies, even the least conservative, exhibit instances of reduced institutions, and of vestiges of institutions which have disappeared.

We will take the second point first, as it can be more briefly dealt with.

1. We know that imitation of the past and respect for tradition and custom are reduced to a minimum in modern societies, especially in the countries of the New World. Even in these recently formed States, however, instance of survival may be found.

In the first place, there are legal and religious survivals of European origin. Jews settling in the United States kept up the practice of circumcision, while Christians introduced the Eucharist. Spencer[1] has shown that forms of greeting are vestiges of a primitive ceremonial demonstrating submission to the omnipotence of others. Then take the Calendar system which is universal; we know that the names of months and days of the weeks are survivals from Polytheistic times, and it seems certain that both circumcision and the celebration of the Mass are true survivals which originated in religious sacrifices. Independently of these imitative survivals vestiges remain in the United States of reduced institutions which were fully functional in the last century.

[1] Essays on Progress.

Take for instance the town meetings of Boston and Newhaven.[1]

When the inhabitants of a town attained to a certain number, the town became a city, and the Assembly of inhabitants was transformed into a Common Council. In some instances, this old system persists though in a reduced condition.

In Boston, which continued to be a town, governed by an Assembly of all the inhabitants up to 1821, the present Charter of the city authorizes the convocation of a town meeting wherever the Mayor and Aldermen consider it advisable; the latter, however, never make use of this privilege.

In Newhaven (Connecticut), the old town meeting continues to exist side by side with the Common Council which was established in 1784, but, Levermore says, " This ancient institution nowadays is a meeting together of a small number of citizens to conduct the business of several thousands. The few people connected with the affairs of the town (which is very poor), meet together to discuss matters in a friendly way, decide what money is required for current expenses, and then adjourn. Not one in seventy of the inhabitants attends these meetings. Few know when they take place, and the papers make brief, if any, mention of them."

2. We have now to show that the simplest societies have undergone modifications, and exhibit

---

[1] Bryce, *The American Commonwealth*, i., pp. 598 and following. London, Macmillan, 1893.

instances of survival. We will take as examples those rudimentary types most nearly approaching to the primitive type,[1] *i.e.* the Veddahs of Ceylon, the Fuegoes of Cape Horn, and the Australian tribes.

(*a*) The Veddahs, who have lived in the jungles of Ceylon for centuries, either as separate families, or in groups of two or three families, appear to have formerly possessed a much more complicated social organization. According to Max Müller, they were not formerly so low in the scale of humanity; he says that their language, if not their blood, betrays their "distant connection with Plato, Newton, and Goethe."

In their language, folk-lore, and clothing, these retain characteristic vestiges of a former condition. Take for instance the carefully observed practice of piercing the ears of children at the age of three or four years, although eventually only a small number of them could wear ornaments in them, others having to be content with small pieces of twig, coiled leaves, or bits of straw.

"This custom," says Deschamps, "is extremely old, and we may suppose—as there is no other signification in it than the prospect of ultimately wearing jewels—that it dates back from a time when the people were not in so low and destitute

[1] "Aggregates formed by a simple repetition of hordes or clans without any such interrelations between them as to form intermediate groups between the whole collection and the individual clans." Durckheim, *les Règles de la méthode sociologique*, Paris, F. Alcan, 1895.

a condition as they are now. Having in more civilized times worn jewels in their ears, the custom of piercing the ears in youth persists, though the jewels may be lacking."[1]

(b) Bridges says that according to a tradition which is probably true, the Fuegoes, until quite recently, submitted their young men to a sort of initiatory trial when they attained to adolescence. They were taken into a hut (the *kina*) set apart for the purpose, and there underwent certain tests, including a rigorous fast. Bridges adds that the *kina* was also the theatre of mysterious and bizarre scenes of very ancient origin, the rôles of which, now relegated to men only, were entirely performed by women. Contrary to Giraud-Teulon who cites these facts as evidence of the former existence of a matriarchy, the fêtes of the *kina* seem to have quite disappeared from among the natives of Orange Bay.

Dr Hyades, however, mentions a survival of the old custom. "The custom is still observed of submitting young girls to a fast at the time of puberty, but this fast is less severe than that already mentioned as undergone by the boys; the same good advice is then given them by their parents, as was formerly bestowed upon the boys in the *Kina*."[2]

[1] Emile Deschamps, *l'Anthropologie*, 1891, t. ii., pp. 297 and following.

[2] *Mission scientifique du cap Horn*, 1882-1883, t. vii. *Anthropologie, Ethnographie*, by P. Hyades and J. Deniker; Paris, Gauthier-Villars, 1891, p. 377.

(c) Recent researches into the family system among the Australian tribes has brought a number of survivals to light. This is especially the case with regard to the careful researches of Fison and Howitt[1] who have shown that, independently of their tribal divisions—which are really territorial groups—the Australians are divided up into clans or sexual groups comprising all the individuals with the same *Kobong*.[2]

The members of these groups are regarded as members of the same family, and may never, under any circumstance, intermarry, under pain of being driven out of the clan and hunted like wild beasts. Sometimes individuals of antagonistic tribes living at several hundred miles' distance from one another and speaking different languages have the same *Kobong*. The law of classes remains active; a captor may not violate a prisoner belonging to his group, but a stranger may enter into relations with the women of another tribe, so long as the tribe belongs to a class related to his own. This system of relationship can only be explained as being a survival from a former period in which all persons with the same *Kobong* belonged to the same group. This is a disputed point,[3] however,

[1] Fison and Howitt, *Kurnaï and Kamilaroï* (*Journal of the Anthropological Institute*, 1884).

[2] "The *Kobong* of a man is the animal or plant, the name of which he bears and reveres as a protecting spirit" (Starcke).

[3] Starcke, *la Famille primitive* (Bibl. sciens. intern., Paris, F. Alcan, 1891, p. 22).

for besides this very likely hypothesis, undoubted survivals remain of intermarriage by groups or sexual groups. In the writings of Fison and Howitt, we find the two following instances of this in two tribes which, according to them, severally represent the highest and lowest in the scale of civilization, among those with which they came in contact.[1]

(d) The tribe called *Kunandaburi* was divided into two exogamous classes: *Mattara* and *Yungo*. Theoretically all the Yungos whether male or female were regarded as the males of the Mattaras, and *vice versâ*. In point of fact, however, only one vestige of the primitive communal marriage remained—the *jus primæ noctis* which was the prerogative of all the contemporaries of the husband belonging to the same group.

(e) The tribe called *Narrinyeri* which represented a more advanced stage of civilization, was equally divided into two sexual groups, but in reality, marriage was strictly individual. One survival remained, however, of the former system. When a man captured an alien bride, all the men of his own generation and belonging to the same group possessed the right of *jus primæ noctis*.

3. We have seen that instances of survival are rare in some countries because modifications are only effected slowly, and in others because changes are effected very quickly and useless institutions

---

[1] Fison and Howitt, *Journal of the Anthropological Institute*, 1882, p. 35.

are eliminated root and branch. It is in countries like England, where modifications are brought about with a due respect for old customs and traditions that ceremonies, institutions and customs exhibit the greatest number of survivals.

§ 2. *Survivals of ancient forms of marriage and of the family in Modern Europe.*

We think we may regard it as proved that all societies exhibit instances of survival, but in order to further demonstrate the universal character of retrogressive evolution, we shall show, by means of a careful study of one particular institution, that vestiges of former institutions are neither rare nor exceptional, taking as examples the various forms of marriage and of the family throughout Modern Europe.

1. *Forms of Marriage.*—From archaic times up to our own, we find that among modern nations, marriage by capture, marriage by purchase, and marriage with the consent of the woman have been successively followed by marriage by simple consent, religious marriage (*in facie Ecclesiæ*) and civil marriage,[1] and that survivals remain of all the forms of marriage anterior to civil marriage.[2]

[1] Paul Viollet, *Histoire du droit civil Français*, p. 424 and following. Paris, Larose & Forcel, 1893.

[2] Westermarck, *The History of Human Marriage*, 1892, p. 418 (when the mode of contracting a marriage altered, the earlier mode, from having been a reality, survived as "ceremony").

Marriage by Capture.—Traces of this early form of marriage, which is still to be met with in certain parts of Bulgaria,[1] are exhibited in the nuptial rites and customs of Ukraine, where the ceremonial is quite a museum of reduced institutions.

Vestiges of what originally was marriage by capture occur too in France and Belgium, and other countries. The nuptial games of Lower Brittany and Chimay have obviously been derived from it.[2]

Marriage by purchase.—The retrogressive evolution of this form of marriage exhibits the following stages :

---

[1] Th. Volvok, *Rites et usuges nuptiaux de l'Ukraine* (*l'Anthropologie*, 1891, p. 169). " In certain parts of Bulgaria (*Kustendil*), the capture of young girls takes place even in the present day and constitutes a form of marriage (*Vlatcheny monny*)."

[2] Monseur, *Bulletin du Folklore*, January and March 1895, *Coutumes*, p. 1 ; *les noces*, p. 18.

"In some villages of Chimay (*Hainault*), when a young man chooses a bride from a neighbouring village, the young people proceed in cavalcade to the home of the bride. The leader of them presents her with a whip and a large cake crowned with a bouquet. The bride then takes up her position on the doorsteps while the bridegroom and his friends pass and repass her at full gallop seeking to dispossess her of the whip with which she lashes at them. All this may be regarded as the last remaining phase of marriage by capture, the bridegroom arriving at the dwelling of the bride with a cortège, being more or less figurative of a marauding expedition."

Baudrillard (*in Séances et travaux de l'Académie des sciences morales*, Jan. 1884, p. 36) connects the customary nuptial games of Lower Brittany with the primitive marriage by capture. In these games the bride hides before going to church, and the bridegroom has to search till he finds her.

1. The payment made was originally a pecuniary compensation to the family of the captured bride. By degrees, actual capture gave place to mere symbolical capture and then the system of compensation became transformed into the purchase system.
2. The price paid for the bride which was originally the property of the whole clan under the name of *wergelt*, became the perquisite of the bride's father.
3. It next became modified into a marriage dowry given to the wife by her husband and the ceremony of purchase became purely symbolical. In the Merovingian period, for instance, the future husband presented the father of the bride with a sou and a denier (marriage *per solidum et denarium*).[1]

Other survivals still exist of these three stages in the evolution of marriage.

(*a*) In Ukraine, where the signification of the purchase system was purely one of the *vira* or compensation (*wergelt*), the bridegroom has to give a present to each member of the bride's family; this is the custom too in the valleys of the Caucasus where the

---

[1] Vanderkindere, *Condition de la femme à l'époque mérovingienne*, p. 12. A. Heussler, *Institutionen des deutschen Privatrechts*, ii., p. 280.

relations make a further demand of payment in money.[1]

(b) The custom of purchasing the bride from the father still exists—although with no legal recognition among the Ossetes and in certain Russian villages.

(c) Lastly, the marriage *per solidum et denarium*. Traces of this are left even in modern France. This particular purchase signified the payment of thirteen silver deniers according to a certain value of the sou.

The introduction of these thirteen deniers into the ceremony of marriage can be traced through the middle ages up to the marriage of Louis XVI. when they still appeared. In some parts of France such as Dijon, Bordeaux and Barrois, they may still be met with even nowadays. With the exception of this figurative number of thirteen, characteristic of the primitive origin of the ceremony, this form of marriage has undergone such changes as almost to entirely obliterate its primitive character. This is so, too, with other survivals of the system of marriage by purchase—they have come to be only intelligible by means of the comparative method, or by a knowledge of their historical antecedents.[2]

---

[1] Volvok, Journal *l'Anthropologie*, 1892, p. 579. Kowalevsky, *Droit coutumier osséticu*, p. 176.

[2] See Paul Viollet, *Histoire du droit civil français*; Paris, 1893, p. 403.

Marriage fairs—now transformed into kermesses, and still held at Lierre in Belgium and other places—appear to have been originally regular markets for the purchase of young women.[1]

The custom of offering the wrong young woman to the bridegroom as his future wife—a custom still in vogue in the department of Landes in France — formed part of the nuptial ceremonies among the ancient Hindoos, and is probably a vestige of the tricks which were played upon the bridegroom after the purchase of the bride. Several vestiges of this kind may be noticed among the customs of modern peasants. It is by no means infrequent for a wife to be sold by her husband, on the principle that what has been bought may fairly be sold, and the transaction rendered legally binding by being drawn up on stamped paper. This is merely a survival of the old system of marriage by purchase.

[1] "The second Sunday of this fair, which commences on the Sunday after All Saint's Day, is called the *veersensmarkt* (Heifer-fair or market), and the third Sunday is called the *brullenmerkt*, a name derived from the word *brul*, and signifying a heifer lowing noisily. The *veersensmerkt* is the day especially set apart for the young girls who attend the fair to find husbands. The *brullenmerkt* is the day dedicated to older women as a sort of forlorn hope for those who have hitherto failed to get married. No one seems to know when these "markets" first came to be held. I am more than sixty years old, and they were old when I was young, with this difference only, that in my youth, a young girl who respected herself would not have been seen at the *brullenmerkt*, which is not the case nowadays." (Taken from a letter from the Secretary of the Commune at Lierre.)

Marriage by consent of both parties.—In French civil marriages of the present day, traces remain of the two other forms of marriage by common consent which preceded it.

(a) The *marriage by simple agreement*, which held good throughout Christian Europe until the Council of Trent, is still valid in Scotland. Gretna Green marriages (Gretna Green being a village situated on the Border, near Carlisle) were notorious. In accordance with an old custom, the blacksmith of Gretna Green kept the register of these marriages, and the union was contracted in his presence.[1]

After the Council of Trent, marriages in *facie Ecclesiæ* came to be alone recognized by the Church: all marriage to be valid must be sanctioned (if not actually celebrated) by the parish priest of one contracting party, in the presence of one or more witnesses. The legislation of the Council of Trent became French law after the Ordinance of Blois in 1574, until the law was revised in September 1792, when civil marriage was definitely established in France.

[1] Viollet, *Histoire du droit civil français*, p. 428.

"These marriages, about which everyone has heard, were not an invention of Scotch law. Like most legal curiosities, they find an explanation in survival from a former condition of things, persisting in a singularly original and peculiar form.

"The blacksmith and his register were not necessary in themselves to contract a marriage; they merely constituted evidence that it had taken place. In 1804 a Scotch man and woman merely declared themselves man and wife in writing, and in 1811 they were recognized as legally married by Scotch law."

From that time forth, religious marriage may be regarded in the light of a survival, having lost all legal importance, while civil marriages are greatly on the increase. Priests, then, the "groomsmen" of Lower Brittany, and the blacksmith of Gretna Green may alike be regarded as plain evidences of an institution in process of decline.

II. *The family system.*—Archaic forms of the family still exist, sometimes as mere vestiges, sometimes as exceptional cases, in countries where only separate families are legally recognized.

1. *The Matriarchy.*—Traces of the matriarchy, *i.e.* of the exogamous family of blood relations through the mother, abound among the customs of the inhabitants of the valleys of the Caucasus. Kowelevsky, in his book on the customs of the Ossetes, has dealt with this subject. The vestiges remaining even in France of marriage by capture and the prohibition of certain marriages in Montenegro, derived doubtless from an exogamous period, are alike survivals of this primitive family system. Further, although both facts and their interpretation are rather doubtful, some authorities regard the *couvade*, a custom which is still in practice among the Basques and also in the Isle of Mark (in Holland), as a vestige of the transition period between the matriarchy and paternal affiliation.[1]

---

[1] See Viollet, *Précis de l'histoire du droit français*, ii. 326, and Giraud-Teulon in *Origines du marriage et de la famille primitive*, Paris, 1884, p. 138 ; and Starcke, *Famille primitive*, p. 49.

2. *The Patriarchy.*—The Patriarchy, which was a system of family community, is still exhibited in the *participanze* of Italy, the *companias de Galicia* of Spain, the *parsonneries* of France, the *Hausgenossenschaften* of Germany, and the *zadrugas* of the Balkan peninsular. Besides these, the family system of to-day, according to Sumner-Maine, affords constant evidences of a *patria protestas* in process of decay and of declining male property rights.

Survivals from the old patriarchal system abound in modern legislation. Take the following examples:

   (*a*) The limitations imposed on a testator with regard to leaving his property away from his family (*Civ. Code*, 213 and following).

   (*b*) The legal opposition to a woman's equality in succession.

      The more barbarous laws, if not actually excluding women from succession, at least excluded their succession to landed estate, in order to keep the family property intact. In France the privilege of sex was maintained in some respects, even among the peasants, up to the close of the old regime, and it still exists in the present day in different degrees in Scandinavia, Russia, Servia, and some of the Swiss Cantons.

   c) The inequality of the sexes as regards conjugal fidelity.

The privileges of the husband over the wife in this matter are survivals from the time when these obligations were wholly on one side.

(d) *Affiliation following on a double marriage.*— In some countries, under the old system, a child could only succeed to the property of his parents if he resided with them (excepting in quite exceptional cases). Hence there arose marriages by exchange. In order to compensate the children, the two families, if each had a son and daughter, exchanged them, and bestowed the rights of one upon the other. Under the new law these rights could not be legally claimed, and yet even in the present century they are not wholly unknown. The last marriage by exchange was probably that mentioned by Dupin[1] which took place at Gacogne (Niévre) in 1839.

In conclusion then, we have found that while the legislative systems of modern races tend to become more and more alike in main principle, we can yet find vestiges, more or less faint and distorted, but quite recognizable, of the different institutions which dominated the earlier conditions of the different races.

[1] Viollet, *Histoire du droit civil français*, p. 491.

PART III

SUMMARY AND CONCLUSIONS

THE examples we have been able to give in the first part of this volume make it plain that degenerative evolution exists everywhere. It must be noticed, however, that biological investigation shows that in the evolution of organs certain parts may disappear completely, but also that in the evolution of organisms certain organs may also disappear. This last phenomenon is most common in embryological development, when it is known as ontological abbreviation.[1] Sometimes it is the adult stage that is suppressed, this being possible by what is called pædogenesis a precocious appearance and ripening of the sexual organs.[2]

---

[1] In Scalpellum Stroïni, a deep sea Cirripede, the *nauplius* stage of the larval life is suppressed at least so far as that is a free swimming larva. Here is at least a physiologically complete suppression of a whole larval stage.

[2] *Axolotl*. Most salamanders pass through a larval stage in which they are aquatic and perform their respiration by means of external gills. In this condition they are incapable of reproduction, and must undergo metamorphosis to secure propagation of the species. In the case of Amblystoma, however, a Mexican salamander, the larval form of which is called the Axolotl, reproduction is possible in the larval stage. Thus most individuals of

Sometimes a degenerative transformation becomes still more complete and wonderful; not only may a larval stage or an adult stage be completely suppressed, but a multicellular organism may even lose its power of dying. It is known that the simplest forms of life are practically immortal: when a microbe like *micrococcus* divides nothing dies, and throughout the whole series of successive divisions the primitive life is preserved. On the other hand, in the case of higher animals such as man there are both mortal somatic cells and reproductive cells which by means of conjugation become practically immortal. The mortality of the somatic cells is evidently an acquisition, an advantage fixed by natural selection; but there exist multicellular organisms evidently derived from creatures which had acquired the division into mortal somatic and immortal reproductive cells and which have lost it since. All the cells of their body are able to avoid death by conjugation. This occurs in many conjugate algae like *spirogyra*

this species do not actually reach the adult stage. According to Boas writing on *Neotenie* in Gegenbaur's *Festschrift*, 1896, this probably happens in the case of all the perennibranchiate urodeles.

Ranunculaceæ. On page 85 we showed that in *Ranunculus aquatilis* there are produced first submerged leaves, and afterwards floating lobed leaves, and that the flowers are produced in the axils of the floating leaves. Some forms of the plant living in deep water produce only lacinated leaves, in the axils of which by a kind of pædogenesis the flowers are produced. In other species (*Ranunculus fluitans* and *R. divaricatus*) the pædogenesis has become definitely established and no floating leaves are formed.

and in some of the Volvocineæ (Stephanosphæra, Eudorina).

Plainly then, the further one examines the facts, the more enlarged becomes the conception of degenerative evolution. It is not confined to unusual, abnormal or pathological cases. Degeneration is not an accident in evolution: it is the obverse of progressive evolution and the necessary complement of every transformation whether anatomical or social.

Whatever transformation may be studied, it will be found that change is always accompanied by an elimination of some parts and that in the interests of the organism as a whole these useless parts gradually degenerate. When a whole organization begins to undergo retrogressive evolution and to decay, it is frequently in the interests of some still larger organization. Individuals or species out of harmony with their surroundings disappear to make room for others. August Comte has shown how death is a progressive agency in the social organization removing the worn-out tissues and leaving room for new and more plastic intelligences. All progress implies necessary eliminations. In the domain of society, those who are victims and who from birth, education, or interests, attach themselves to the decaying institutions naturally see only the degenerative side of the change; but those who regard the process as a whole and do not concentrate their attention upon the injured interests and individual sufferings will see the other side of the movement.

When a social organism is degenerating there is considerable opposition to its complete disappearance, and so as Houzeau has said (see the summary of Book III.) it is to be expected that living and superior civilizations drag behind them a trail of débris from dead civilizations.

BOOK II

THE PATH OF DEGENERATIVE EVOLUTION

PART I

THE SUPPOSED LAW THAT DEGENERATION RETRACES THE STEPS OF PROGRESS

It is a common opinion, supported partly by the etymology of the word, that retrogression is a tracing backwards of progression.

"In the degeneration of organizations fallen out of use," M. A. Lameere has said, "it is to be observed that the structures formed most recently and most specialized are the first to disappear, and that the most fundamental characters are those which persist longest: that in fact degenerative evolution retraces the steps made by progressive evolution. Peculiarities recently acquired, if they become disused, rapidly disappear, while dispositions of a more ancient kind have a persistence almost exactly proportioned to their age.[1]

This supposed biological law of retracement has

[1] A. Lameere, *Esquisse de la Zoologie*, Bruxelles, Rosez, p. 184.

penetrated to psychology and sociology. In 1868 Hughlings Jackson, in the study of certain maladies of the nervous system, had arrived at the conclusion that, "In the degeneration of this system the higher functions, those more complex, specialized and voluntary, disappear more quickly than the lower, simpler, less specialized and more automatic functions."[1]

Starting from this point, and expressing it in terms of physiology, Ribot formulates as follows the law of degeneration of will and memory: "The dissolution of the will occurs in a retrograde fashion, from the more voluntary and complex to the less voluntary and simpler—that is to say, towards automatism."[2]

So also in progressive loss of memory, the degeneration proceeds from the less stable to the more stable. "It begins with recent acquisitions not firmly rooted in the brain, rarely repeated, and so not firmly associated with others, in fact with the least organized parts of memory. It ends with sensory memory which is instinctive, and is deeply rooted in the organism, or is indeed a part of the organism itself."[3]

These retrograde transformations of the nervous centres have their echoes in the modes in which ideas and feelings are expressed. Paul Heger, in

---

[1] Ribot, *Maladies de la Mémoire*, p. 29. Dallemagne (*Dégénérés et Déséquilibrés*), p. 430.
[2] *Maladies de la Volonté*, p. 150. Paris, F. Alcan.
[3] *Maladies de la Mémoire*, p. 94.

particular, has shown this in his investigations into the degeneration of writing and speech.[1]

In a lecture on the evolution of language, delivered at the University of Brussels, he said as follows : " For several years I have studied the degenerative evolution of writing, and I have shown how the writing of the insane resembles that of children. All that I have said with regard to writing applies to speech, and just as drawing lasts longer than writing, so rhythm and music survive after the power of expressing ideas by words has been lost." " The gradual degeneration of speech may be traced in the case of old men who gradually pass down the incline into senility. Study of the speech of such persons shows that the degeneration of their faculty retraces the steps by which it had been progressively acquired."

The labours of Heger were in a field where the social element was important. It is a small step from them to social affairs themselves. A number of authors, including Ferrero, Colajanni and Degreef, base their ideas upon this law of retrogression, which they regard as established and applying to sociological phenomena.[2]

---

[1] *Sur l'Évolution Régressive de l'Écriture chez certains aliénés* (*Bull. de la Société d'Anthropologie de Bruxelles*, v., 1885-1886). *Sur l'Évolution du Langage* (*Revue Universitaire, Bruxelles*, 1892-1893, p. 143).

[2] Degreef, *Le Transformisme Social*. F. Alcan, Paris, 1895, p. 365.

Before describing and discussing the special facts to which they apply their theory, it is necessary to examine its biological foundation.

CHAPTER I

THE PATH OF DEGENERATION IN BIOLOGY

WE have now to consider if the degeneration of organs retraces the steps taken in their progressive evolution. According to Hughlings Jackson and Ribot, in the cases mentioned in the preceding chapter, degeneration proceeds by successive atrophies occurring in the order opposite to that of ontological formation. Is the same order to be found when we compare the degeneration of organs or of individuals with their phylogenetic development? To answer this question, we must employ both morphology and embryology. Using the morphological method, we shall study the reduction of a homologous organ in several species descended from the same type, and compare the different stages of reduction with the different stages of phylogenetic development of the organ.[1]

---

[1] In this investigation it will be necessary to compare absolutely identical organs—for instance, not to compare the pineal eye with the paired eyes. It will be necessary also to choose animals of common parentage—to avoid, for instance, comparing a Vertebrate eye with a Crustacean eye.

Using the method of embryology and the principle, so far as it can be followed, that ontogeny recapitulates phylogeny, we shall investigate the mode of origin of some reduced structures. We shall thus learn if the organs in degenerating resume any of their ancestral stages.

Reduced organs may appear in two different forms. They may be atrophied after having reached a more or less complete development, and in this case we shall have to compare the course of the atrophy with the course of the development. They may be rudimentary, that is to say, their development may have been arrested at a given point, the adult state never being attained. In this case, so far as ontogeny repeats phylogeny, the arrest of these organs at different stages in different species should furnish a series with greater or smaller lacunæ, but a series which will be the reversed series of the original stages in phylogeny.

SECTION I.

The path of degeneration in animals.

1. *Morphology and Embryology. The law of recapitulation.*—It has been so often repeated that the individual development of an organ is a résumé more or less compressed of its historical evolution, that people are apt to attempt too exact an application

of this principle to every individual case. Such an exact application is, however, impossible. Every living organism is plastic, and in its development presents individual variations which serve as material for the operation of natural selection. In consequence, the recapitulation cannot be more than a repetition more or less vague of the essential phases of phylogeny.[1]

Moreover, there is nothing inevitable in the law of recapitulation, for most plants develop directly.

With these limitations, however, we may state that among animals, the ontogeny usually repeats in a modified fashion the main ancestral stages. This is certainly the case when we compare the development of the brain of man with the probable ancestral stages as displayed in the series of vertebrates.

---

[1] Lang, *Anatomie comparée*. As throughout the whole course of time, adaptation, that is to say, the preservation of what is most useful in the struggle for existence, is a force modifying heredity, it is plain that a species instead of resting stable must change. According to its circumstances, moreover, the successive stages in the ontogeny of a creature are under the influence of conditions different from those that affected the corresponding ancestral stages. We shall call the process of embryology palingenetic so far as it is based upon inherited legacies, and cœnogenetic so far as it is modified by adaptation.

Baldwin in his Treatise on *Mental development in the child and in the race* (London, 1895), also shows that the development of the individual is not an exact repetition of ancestral stages. The development of the child exhibits "short cuts" and phases of direct development due to adaptation and destroying the exactness of the parallel with phylogeny.

In Fish and Batrachia.	The cerebral hemispheres do not cover the region of the third ventricle from which the eyes arise (thalamencephalon).	In the human embryo (fig. 61, A) of the seventh week.	Same aspect.
In Reptiles.	The hemispheres cover the thalamenaphalon but leave uncovered the region of the optic lobes (mesencephalon).	In the human embryo (fig. 61, B) of the middle of the third month.	Same aspect.
In mammals.	The hemispheres cover the thalamencephalon, the mesencephalon, sometimes the metencephalon (cerebellum and medulla), and the olfactory lobes.	In the human embryo (fig. 61, C) of the fifth month.	Same aspect.
In some mammals even of higher orders (*e. g.* some Hapalidæ).	The hemispheres are smooth.	In the human embryo (fig. 61, D) of the midddle of the fifth month.	Same aspect.

Within such limits, the law of recapitulation may be applied, and the embryonic history of an individual may be considered roundly as a repetition of the essential phases of its ancestral history. We have now to consider how far a reduction by atrophy or by arrest represents a retracing of steps in evolution (fig. 61).

From this point of view, we may study the

degeneration of the pineal eye in the slow-worm, and in a series of lizards.

The pineal or median eye in the slow worm and

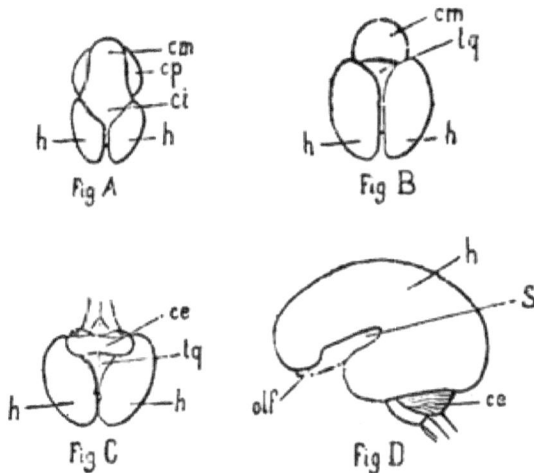

FIG. 61.—*A*, Brain of a human embryo of seven weeks; *h*, cerebral hemispheres; *ci*, intermediate brain or thalamencephalon; *cm*, mid-brain; *cp*, hind-brain. *B*, Brain of a human embryo about the beginning of the third month; *h*, cerebral hemispheres; *lq*, region of the corpora quadrigemina; *cm*, mid-brain. *C*, Brain of a human embryo at the middle of the third month; *h*, cerebral hemispheres; *lq*, corpora quadrigemina; *ce*, cerebellum. *D*, Human brain of the fifth embryonic month; *h*, cerebral hemispheres; *olf*, olfactory lobes; *S*, fissure of Sylvius; *ce*, cerebellum. (After Mihalkovics, *Entwickelungsgeschichte des Gehirns*. Leipzig, 1877.)

the lizard passes through the following stages in its individual development.[1]

(1) Formation of a hollow outgrowth from the roof of the third ventricle of the brain (fig. 62, D).

[1] P. Francotte, *Recherches sur le développement de l'épiphyse.* (*Thèse présentée à la Faculté de médecine de Bruxelles.*) *Arch. de Biologie*, 1888.

(2) This little sac elongates, changes its direction,

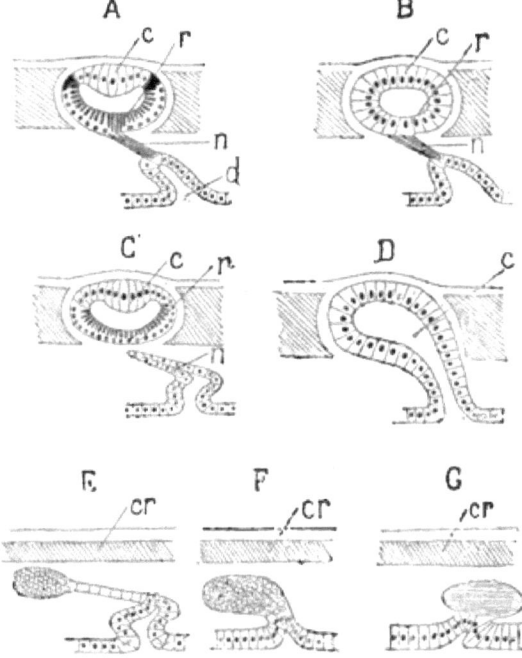

FIG. 62.—Diagram indicating the progressive evolution and the degeneration of the pineal eye.
A. Perfect pineal eye, as found in the slow-worm before birth, or in the adult *Sphenodon* (Hatteria) ; *c*, lens ; *r*, retina ; *n*, optic nerve ; *d*, divertialum of the thalamencephalon. B. Pineal eye in first stage of degeneration as it exists in *Chamæleo* and as it was in the slow-worm before stage A. The lens (*c*), and the retina (*r*), are not differentiated. C. Pineal eye in the degenerate form found in *Calotes* and *Leiodera*; *c*, lens; *r*, retina; *n*, optic nerve in fatty degeneration. D. Very degenerate pineal eye as in *Cyclodus* and like the earliest stage in the slow-worm ; there is no differentiation of the divertialum from the thalamencephalon. E, F, G. Other modes of degeneration of the pineal eye. The eye lies within the skull and there is no parietal foramen ; *cr*, cranial membranes; E. *Ceratophora*. F. Birds ; *g*, mammals. (After Baldwin Spencer.)

and becomes divided into a proximal and distal portion. The cells lining the distal part, that farthest from the brain, become differentiated into the cells which will form the lens, and the cells which will form the retina.

(3) The distal part becomes specialized, the lens, the retina, and the stalk of the optic nerve are mapped out.

(4) The lens, the retina, and the optic nerve become fully formed (fig. 62, A).

At this stage the third eye has reached its limit of development.

There is a well-formed retina connected with the brain by a special optic nerve. The organ projects strongly from the surface of the head, but from this point, owing to the development of the cerebral hemispheres, degeneration begins. The nerve (fig. 52, C), becomes broken and fatty, and pigmentary degeneration occurs in it. At the same time, the pineal eye having become useless or even harmful to the animal possessed of it, before the power of receiving perceptions of light has been lost, and before the organ has been far reduced by phylogenetic destruction, a veil of black pigment is formed over it, completely shutting it off from the outer light. The nerve disappears completely before birth, its degenerate cells becoming lost in the mesoblastic skeletal tissue of that region. At the time of birth the whole eye is enclosed in a thick membrane which isolates it. The deposition of pigment has destroyed any functional activity in the lens and the retina, but these parts none the less retain traces of a complicated structure recalling their condition when functional.

In the Rhynchocephala and Lizards examined by W. B. Spencer,[1] there is to be found a series of types representing the various stages of the degeneration of the eye in the slow-worm.

In the type *Sphenodon* (*Hatteria*, fig. 62, A), the organ in the adult is in the complete form. The eye possesses a lens (c), a retina, (r), with complicated histological structure. A nerve (n) places the retina in communication with the brain.

In *Chamæleo* (fig. 62, B), the degradation of the organ has reached the following stage: the epiphysis has a distal portion corresponding to the eye, but the histological differentiation of this is incomplete, neither the retina nor the lens being distinct. Nervous fibres connect this with the proximal portion which is hollow and in communication with the brain. It thus represents the second stage in the formation of the eye in the slow-worm.

In the types *Leiodera* and *Calotes* (fig. 62, C), the chief degeneration is in the optic nerve, which has partially disappeared, and no longer connects the eye with the brain. The eye itself is not quite

---

[1] W. B. Spencer, *On the presence and structure of the pineal eye in Lacertilia* (*Quarterly Journal of Microscopical Science*, 1886).

so degenerate: the retina has not a complex structure, but both retina and lens are present.

In the type *Cyclodus* (fig. 62, D), the degeneration is still greater: the epiphysis is a vesicle attached to the thalamencephalon. The walls of this vesicle show only the smallest symptom of primitive differentiation into lens and retina. The proximal part remains hollow, and shows no trace of differentiation at all.

This, however, is not the only fashion in which degeneration of the pineal eye proceeds. In another series of creatures it retains its connection with the thalamencephalon, but remains inside the skull. The parietal foramen closes, thus completely shutting off the eye from the light; the eye becomes useless, degenerates, and the optic nerve loses its function as a conducting channel, *Ceratophora* (fig. 62, E). The pineal organ then becomes a degenerate structure in which it is exceedingly difficult to see traces of its original condition, and which is usually marked by an abundance of blood-vessels: Birds (fig. 62, F), Mammals (fig. 62, G).

Thus, the degeneration of the pineal eye shows that the optic nerve, the last organ to be completed, is the first to disappear.[1]

In studying the degeneration of the pineal eye,

---

[1] The same happens in the case of the ordinary paired eyes. In the degenerate eyes of the mole, the optic nerve is more reduced than are the other structures.

we have been seen the path of progressive evolution being retraced, at least in the case of these reptiles. The present state of knowledge does not permit an extension of the enquiry to the conditions found in higher animals. However, according to Flesch, traces of a sensory epithelium have been found in the pineal body of man, the horse, the sheep, and the bat: that is to say, that in these creatures too the oldest parts of the structure have resisted degeneration longest. We cannot refrain from the conclusion that in this series degeneration retraces to a large extent the steps of original advance.

2. *Degeneration of the organs of sight in deep-sea Crustacea.*—We cannot however establish the conclusion of the last paragraph as a general principle.

The fauna of the deep sea includes a large number of Crustacea, and in these the eyes, which are relatively useless, are often degenerate. The course of the degeneration is generally definite, and of all the structural parts the most long-lived are the eye-stalks, although we know that these are a recent formation. A number of examples chosen from Decapod Crustaceans, which are specially abundant, will illustrate this point.[1]

Nephropsis, which lives in the Atlantic and Indian Oceans at moderate depths, is a relative of the Lobster. The optic stalk is short and carries

---

[1] See Pelseneer, *l'Exploration zoologique des mers profondes* (*Conférences Universitaire de Bruxelles*). 2 Année, 1890.

a rudimentary eye which has neither pigment nor cornea and is coloured like the general surface of the body.

Eryonicus (fig. 63) belongs to the same group, and comes from the region of Saint-Thomas in the Antilles, where it lives at a depth of about 825 yards. This animal has a reduced optic stalk, but at the extremity of this, where in littoral forms the eye is borne, there is only a depression as if the eye had been carefully scooped out.

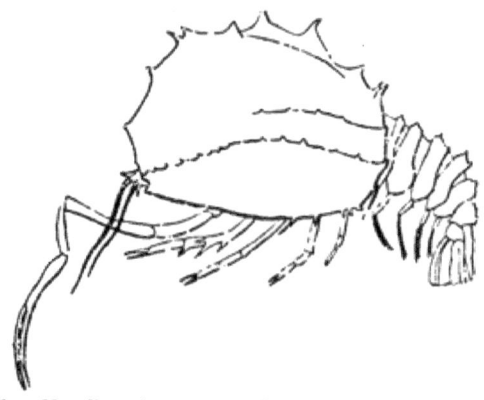

FIG. 63. *Eryonicus cæcus.* Bate? (After W. Faxon, *The Stalk-eyed Crustacea*, Mem. of Mus. of Comp. Zool. Harvard College, vol. xviii., 1895.)

Willemœsia (fig. 64), a relative of the marine crayfish and an inhabitant of the Atlantic at a depth of about 3500 yards, is completely devoid of eyes in the adult condition, although it possesses them in the larval stage.

Scolophthalmus (fig. 65), which lives down to 4000 yards, is quite devoid of eyes, but possesses eye-stalks which terminate in spines.

It seems, then, that different species of deep-sea Crustacea may present different degrees of degeneration of the eye. One species in itself exhibits all

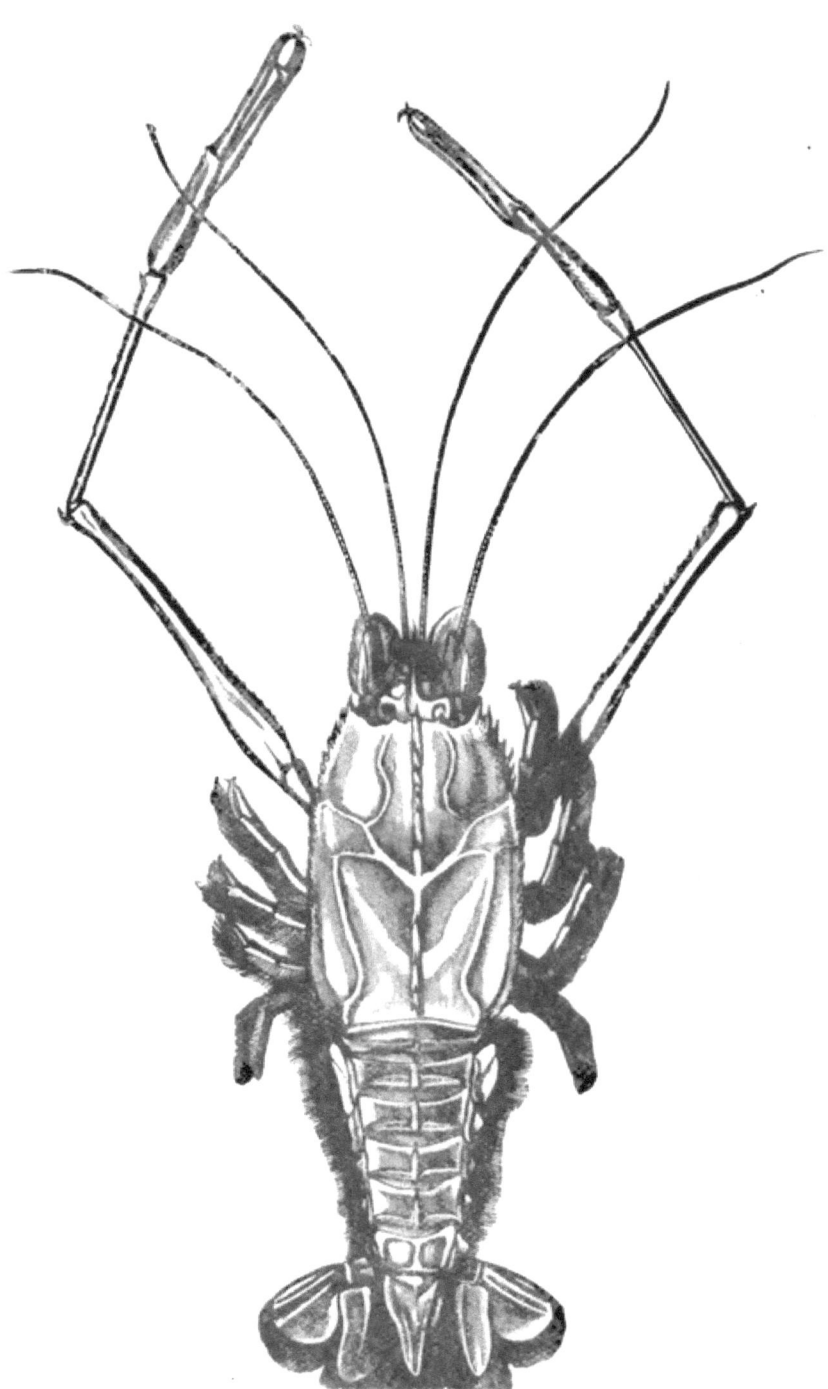

Fig. 64.—*Willemœsia.* (After Pelsencer.)

grades of degradation according to the depth at which it lives. This creature—*Cymonomus*—which, when near the surface, has fully formed eyes upon movable stalks, at a depth of a few hundred yards exhibits movable stalks without eyes; and at 1500 yards the stalks are fixed and end in spines.

Isopod Crustacea, which live in the deep sea, present similarly degenerate eyes. Many are blind

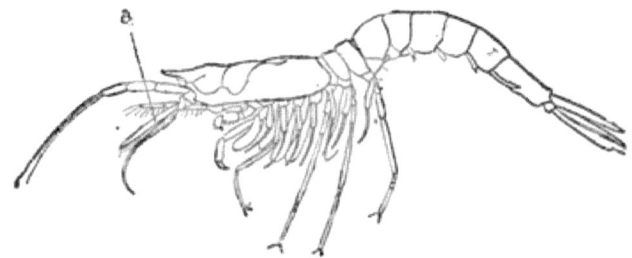

FIG. 65.—*Scolophthalmus lucifugus*, FAX.
a, optic peduncle transformed to a spine. (After W. Faxon, *The Stalk-eyed Crustacea*, Mem. of Mus. of Comp. Zool. Harvard College, vol. xviii., 1895.)

and display all kinds of optic degeneration. *Nocsa*, for instance, simply has eyes devoid of pigment. Thus, in abysmal Crustacea, the degeneration of the eyes is in no sense a retracing of developmental stages.

Another instance chosen from examples of the atrophy of organs in individuals, shows that the supposed law of retracing cannot be made universal.

4. *Atrophy of the branchial vessels in man.*—Examination of a human embryo of about three

weeks old shows the presence of a series of slits on the sides of the neck, the slits not being parallel, but converging towards the ventral surface. Between these slits are swellings, or pads, which pass up towards the dorsal surface and appear like the beginnings of hoops or ribs enclosing the visceral cavity; the elevations are the branchial arches, the slits are the gill-slits.

In the human embryo (fig. 66, A), as in fish, these slits appear from above downwards, and as they are formed, the corresponding blood-vessels arise.

These vessels, or aortic arches, arise from a ventral aorta (*a*.) which gives off six lateral branches (*c*.) at each side. These lateral branches run up between the gill-slits and form two main trunks on the dorsal side which converge to form the descending aorta (*ad*.).

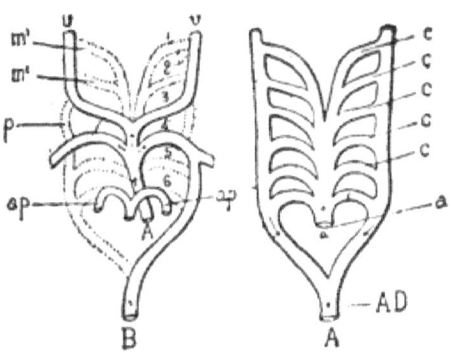

FIG. 66.—Diagram of branchial arches in mammals. A. Embryonic stage. *a*, aorta; *c*, aortic arches; *AD*, dorsal aorta. B. Adult stage. The parts represented by dotted lines have degenerated. *A*, aorta; *v*, carotid; *ap*, pulmonary artery.

In man the branchial arches are transformed, parts of them entering into the structure of the face, and during the transformation parts of the aortic arches atrophy (fig. 66, B). But the order of this atrophy does not correspond in any way to the order of the formation of the vessels.

The median parts of the anterior two lateral branches (m^1 and m^2) disappear, and the vertical parts remain as the internal and external carotid vessels. The vertical piece which joined the posterior parts of the third and fourth arches disappears: the internal and external carotids thus acquire a stem of their own. The parts of the fourth arch remain; the fifth arch disappears at each side, and the sixth arch forms the pulmonary artery ($ap.$).

Thus the degeneration of these vessels represents in no way whatever a retracing of their developmental history. All that occurs is that the useless parts disappear and the useful parts persist. A comparative study of this example would only enforce our conclusion.

In ontogeny the neurapophyses are more ancient than the vertebral centres. None the less, as we have already seen, the examination of any vertebral column from head toward tail shows a gradual disappearance of all parts except the centra, although the centra are the last to be formed.

SECTION II.

The path of degeneration in plants.

1. *Rarity of cases of recapitulation in the organogeny of leaves.*—We have already said that

recapitulation seldom occurs in plants, the development of the whole and of its organs being usually direct. When it does occur, it is generally limited to characters coming from comparatively recent ancestors and not even in the most transitory form entering into the formation of the fundamental parts of the plant.[1]

[1] The rarity of recapitulation among vegetables is the result partly of their fixed condition in the soil, and partly of the more rigid nature of their cells.

The immobility of a plant forces the adult to live in the same place as the embryo. Among animals, on the other hand, it frequently happens that the young pursue a manner of life different from that of the adult and resembling that of the ancestor. Young Cirrepedes are vagrant and have the same needs and use the same organs as other vagrant Crustacea; larval frogs inhabit the water like their fish-like ancestors. In plants there is nothing similar; all the aquatic flowering plants are derived from terrestrial ancestors, but if at the beginning of their existence these aquatic plants were to bear leaves adapted to aerial life they would ensure their own destruction. The exceedingly rare ancestral traits to be found in a few species are naturally of a kind not to incommode their possessors. It is improbable that these are a legacy from distant ancestors; they would not have been spared by natural selection had they not come from ancestors of very much the same habit. The absence of locomotion in plants has also produced a greater adaptability than among animals. Animals, when conditions are unfavourable can remove in search of more suitable localities, plants being fixed in the soil must become modified or perish. Plants, therefore, offer numerous cases of individual adaptation. We do not know if these adaptations are transmitted by heredity, but natural selection has at least secured the widest range of plasticity. Thus plants rapidly rid themselves of ancestral legacies which have become useless.

The transitory organs of animals are employed for the service

In consequence, vegetable embryology is of little use for investigation of the supposed backward path of degeneration, for the rudimentary or reduced organs of plants do not generally represent ancestral stages.

The seedling of *Lathyrus tenuifolius* (fig. 67), a vetch, possesses rudimentary organs which cannot be ancestral stages as their development is direct. In this plant a whole series of leaves are formed between those arising at germination and the adult leaves. This intermediate series displays many arrests of development.

The adult leaf has a pair of stipules, foliage leaflets, and tendrils (fig. 67, J). The leaves just before these, have a pair of stipules (fig. 67, I), which are absent in the leaves next before (fig. 67, H). Still earlier leaves are produced with fewer leaflets and tendrils (fig. 67, D-G), leaves without leaflets and with a single tendril (fig. 67, C), and leaves entirely without tendrils (fig. 67, B). Lastly, at germination very rudimentary leaves are pro-

---

of the whole body, the branchial arches of mammals are employed in the formation of important parts of the head and neck. The tail of the tadpole is reabsorbed by phagocytes and its substance used for the nutrition of the body. In the case of plants, such occurrences are rare and limited; the cells are enclosed in a rigid wall which resists displacement or alteration; the protoplasmic contents may be absorbed and used as nutritive material by another part, but the cellulose cell-wall remains. A useless organ can be eliminated only at the expense of loss of material. J. Massart, *La Récapitulation et l'Innovation en embryologie végétale* (*Bull. Soc. Roy. Bot. Belg.*, t. xxxiii., p. 150, 1894).

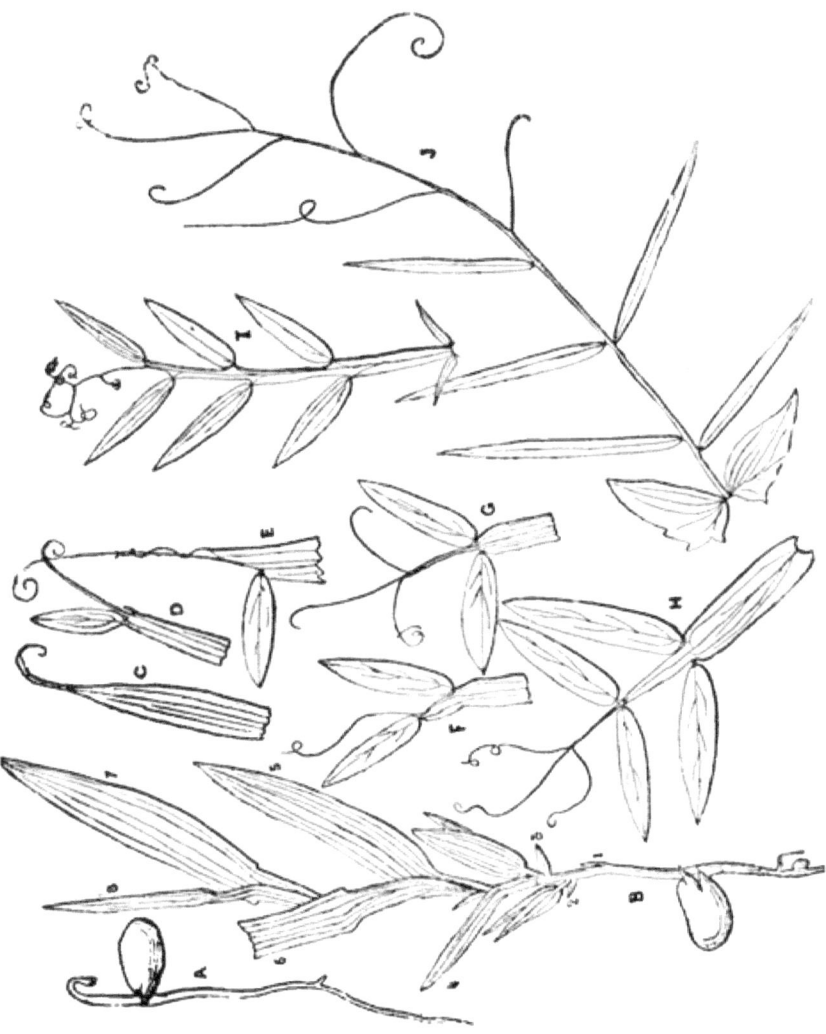

FIG. 67.—*Lathyrus tenuifolius.*
A, B, seedlings in two stages of growth. C to J, different forms of successive leaves.

duced, the sole function of which is to protect the axillary bud (fig. 67, B, 1, 2). We have now to show that this series is by no means a retracing of ancestral stages. All, or nearly all, the Papilionaceous plants have stipulate leaves; this shows that the exstipulate leaves of *L. tenuifolius* do not represent an ancestral stage. Moreover, before the acquisition of tendrils, vetches had a terminal leaflet (see *Vicia Pyrenaica*, fig. 44); none of the reduced leaves in *L. tenuifolius* reproduce this stage; moreover the simple leaves without leaflets do not represent an ancestral condition; the winged petiole is not an ancestral character.

Fig. 68.—Seedling *Vicia* of *monanthos*.

The primary leaves of another vetch *Vicia monanthos* confirm our conclusion. In this case the first few leaves formed in the seedling remain rudimentary, aud serve only to protect the axillary buds (fig. 68, leaves 1 and 2). Contrasted with the condition in *L. tenuifolius*, although all the vetches had probably a common ancestor, these primary leaves have three little projections, the two lateral of which are reduced stipules.

We do not know of any vegetable example of recapitulation in the case of an organ reduced by arrest of development. Plants exist, however, which after having produced leaves of typical structure begin to produce leaves the development of which remains arrested. Such a plant is the *Acacia* which bears phyllodes. At first the leaves are like those of other *Acacias*; next it bears leaves, the blades of which are rudimentary; and finally leaves with normal stipules, but with no trace of lateral leaflets on the petiole.

In this case the ancestral conditions are known and are quite different.[1]

2. *Organogeny of flowers.*—What we have said about leaves applies to the organogeny of flowers. Here also in the cases of atrophy produced by arrest of development there is no indication of

[1] Many other reduced plant organs might be instanced, such as the stipules of *Sambucus* or the teeth of the calyx, in many Compositæ and Umbelliferæ. After their formation such organs grow very slowly and exhibit no trace of recapitulation. This happens with the leaves of *Sempervivum* (fig. 49). Without doubt these leaves are derived from leaves normally divided into hypopodium and epipodium, but they show no trace of this division.

ancestral stages, and thus there is no evidence that degeneration retraces the path of progressive evolution.

In the cauliflower (*Brassica oleracea*, var. *Botrytis*), a cultivated variety, the inflorescence branches exuberantly; most of the flowers produced on these branches are arrested in their development. Of the immense number of flowers produced on each plant only a few attain sexual maturity and produce seeds; the others abort at different stages. Most of these remain in a very primitive stage and do not develop sepals; but, scattered among them, may be found more fully developed flowers, so that the same plant presents almost every possible stage of flower development.

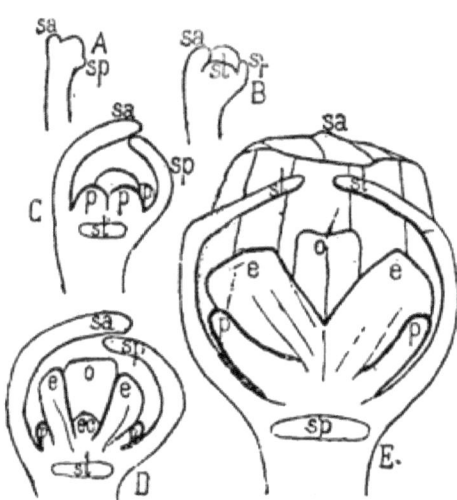

FIG. 69.—Development of flowers of the Cauliflower. *sa*, anterior sepal; *sp*, posterior sepal; *sl*, lateral sepal; *p*, petal; *e*, long stamen; *ec*, short stamen; *o*, ovary. A, very young flower with only rudiments of two sepals; B, flower with rudiments of four sepals; C, older flower with rudiments of petals; D, still older flower with rudiments of stamens and ovary. (The flowers A—D are seen from the side. In the figures C and D the lateral sepal has been removed.) E, flower seen from behind; the posterior sepal has been removed.

The organogeny of these flowers is quite like

that of other Cruciferæ. If in the cauliflower the arrested flowers corresponded to different ancestral stages the case would be striking; but this does not occur. Primitively the flower contained only the stamens and pistils, the essential organs of reproduction. But we see that the cauliflower produces first the anterior and posterior sepals (fig. 69, A), then the lateral sepals (fig. 69, B), then successively the petals (fig. 69, C), the four larger stamens (fig. 69, D), and the two shorter stamens (fig. 69, E). Moreover, the flowers display a special readiness to the suppression of certain parts such as the petals, as in the flower D (fig. 69).

3. *Progressive degeneration of the prothallus in phanerogams.*—Although embryology gives us few examples, morphology proves clearly enough that in plant degeneration there is no return to ancestral types.

This appears clearly from a comparison of the progressive evolution of the prothallus in cryptogams with its degeneration in phanerogams.

Terrestrial vegetation has been derived entirely from aquatic life. The Bryophyta (mosses and liver-worts), the Vascular Cryptogams (ferns), and the flowering plants all have sprung from aquatic algæ probably not very different from *Colcochæte*. Such aquatic forms are reproduced by means of true ova and spermatozoa. The terrestrial plants which were derived from them had also spermatozoa with vibratile locomotor hairs and impregnation took place in a fluid. A special organ, the archegonium, was developed; this contained the ovum and made the approach of the spermatozoa more easy.

For the present purpose we may omit consideration of the mosses and liver-worts as it is improbable that they are in the line of ancestry of the flowering plants. It is necessary only to say that in them, while a single egg is produced in the archegonium a large number of spermatozoa are produced in the antheridium.

The same condition is found among the fern-like plants, but in their case, owing to the development of special channels for the passage of nutritive materials, it is possible for a much greater size to be reached. In these circumstances it would be unlikely that the spermatozoa should find at the summit of a comparatively lofty plant the drop of water necessary for the task of fertilization.[1] Accordingly the sexual cells are produced on prothalli, which hardly reach above the surface of the ground, and are in a favourable position for the necessary moisture.

In the less specialized ferns (fig. 70) a single prothallus bears both male and female organs (antheridia and archegonia). In the Equisetums

[1] The smaller forms like the Selaginelliadæ and the club-mosses were represented in the past by plants of much greater size, as is seen from fossil remains, and it is probable that the modern forms have descended from these giants.

the prothalli are unisexual, but the spores from which they are produced are alike. Finally, in the allies of Selaginella which are the Cryptogams most nearly approaching flowering plants, the prothalli which bear male organs are quite distinct from the prothalli which bear female organs; moreover, the spores which give rise to the two kinds of prothalli are quite different. A small number of large spores arise in special sporangia, termed macrosporangia, and it is from these that the female prothalli arise. A larger number of smaller spores are produced in microsporangia and these microspores produce the male prothalli.

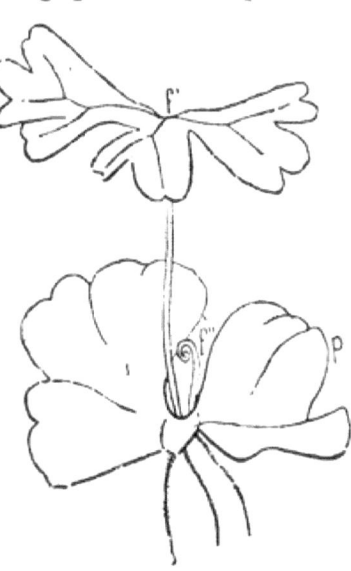

FIG. 70.—Seedling of Fern (*Asplenium Filix-fœmina*. The prothallus (*p*) bears antheridia at its edges and towards the centre an archegonium which has been fertilized. The young seedling developed from the fertilized ovum has already given rise to roots and to two leaves.

The spores of ferns are very small and contain only the nutritive matter required for germination; the spores of Selaginella, even the microspores, are very much larger and contain a large supply of reserve material at the expense of which the development of the prothallus takes place.

We may give a brief account of the formation of the sexual cells in Selaginella. The first stages in the appearance of microsporangia and macrosporangia are identical. Later on in the one case each of the spore mother-cells breaks up into four microspores, in the other case only one mother-cell breaks up into four macrospores, the others disappear. On germination the macrospore becomes a fairly large prothallus with several archegonia: the microspore forms only one cell as the representative of the prothallus, the rest of the structure becoming the antheridium.

In Phanerogams (flowering plants) a drop of water is no longer necessary for fertilization. The spermatozoon reaches the ovum, not by swimming movements, but by a kind of growth. This difference has resulted in profound modifications. Fertilization may occur in the air, and it is no longer necessary for the macrospore to become detached: it remains fixed on the individual from which it arose and there proceeds to develop. In the Gymnosperms the archegonia are fully formed in a prothallus of considerable size, but the Angiosperms which are more specialized, retain only traces of these structures. The microspores (pollen grains) exactly as in Selaginella are produced in great numbers, but when they leave the parent plant they develop only if they reach suitable conditions. In germination there is shut off, as in Selaginella, a single cell to represent the prothallus; later on in a few Gymnosperms it undergoes a few divisions representing the formation of the antheridium. In Angiosperms the reduction has proceeded further, and each pollen grain besides the prothallus cell produces only one or at most two reproductive nuclei. In all Phanerogams the reproductive nucleus reaches the ovum by being carried in a long tube which grows out from the pollen grain at the expense of nutriment derived from the female tissues.

Thus, the prothallus may be traced through marked stages of reduction from the condition in ferns, through Selaginella to Gymnosperms and higher flowering plants. But these phases of reduction are by no means to be regarded as repetitions of ancestral conditions.

In conclusion, it is plain that we cannot assert as a general law, that degeneration retraces the steps of evolution. In the vegetable kingdom no facts support such a conclusion. In certain special cases in the animal kingdom, the most recently developed structures are the first to disappear when atrophy sets in, but this proves no more than that these particular structures happened to be less stable, and to offer less resistance. It is quite impossible to make such facts support the generally held view, that degeneration is a kind of inverse of evolution.

It is very seldom that a living apparatus with complex functions loses all of them, but usually preserves one or other; for instance, the leaves of parasitic flowering plants continue to serve as protectors of the buds, and degeneration does not affect the part that has remained functional. It is highly probable that the original function of leaves was assimilative rather than protective, and yet here it is the later function that is retained.

Even when all function is lost, and the whole structure degenerates, there is no reason why the degeneration should retrace the evolution. In the case of the atrophy of an organ in an individual especially in such pathological instances as those mentioned by Ribot, it may be that the latest formed parts are the most fragile, and the most ready to disappear; but the path of atrophy is quite different in the case of the gradual reduction of an organ in a species. When an organ becomes useless to a species, as in the case of the eyes of deep-sea Crustacea, the only thing that matters to the species is that it may be got rid of. Any individual variation tending towards reduction will be of advantage, and may be retained by natural selection. There is no reason to suppose that such individual variations appear in any inverse order; in fact we do not know that the appearance of variations follows any law at all. Perhaps the apparent inverse order of the degeneration of the pineal eye in lizards may be explained from the fact that the most recently acquired characters are frequently the most variable.

However, even when the most recent organs disappear first, we cannot speak safely of a law of degeneration inverse to evolution. In the actual evolution of many organs, parts have appeared and then completely disappeared. If degeneration were a true inverse of evolution, it would be necessary that such lost parts should reappear only to disappear again. Such observations apply both to ontogeny and phylogeny.

CHAPTER II

THE PATH OF DEGENERATION IN SOCIOLOGY

§ 1. *Investigation of facts.*

WE have now to see if degenerative evolution in social matters retraces the steps of progressive evolution.

In the first place the question cannot be even entertained with regard to some cases, and this for a general reason which will be dealt with later on.

When a complex institution—such as a commercial society or an administrative organization—becomes useless and ceases to be functional, it usually disappears either by voluntary dissolution or else it is legally suppressed. In either case there is no slow retrogressive degeneration retracing inversely the steps of progress, for all the parts cease to exist simultaneously. If certain parts of the suppressed institution are allowed to persist, these are by no means necessarily the oldest parts, but quite the contrary.

When, for instance, the Provincial States of Dauphiny and Normandy were suppressed by the French monarchy, only the titles with their corresponding emoluments were allowed to remain, and they were obviously of more recent origin than the States themselves.

It must be borne in mind that all the parts of an institution rarely become simultaneously useless and non-functional. Those which retain their utility longest are by no means always the most ancient in origin.

English sheriffs have gradually become of less and less functional importance, and now fulfil no other *rôle* than that of presiding over elections and accompanying the judges when on circuit. Both of these functions have been acquired recently compared with all those which the sheriff discharged in the days when the care and protection of the whole county practically devolved upon him.

The question then of the pathway of degeneration only arises in those cases where the same cause of dissolution simultaneously affects all parts of the institution, and where, without sudden interruption, degeneration is effected slowly but surely through many successive stages. This, of course, happens in the degenerative evolution of individual societies or institutions, and not in the disappearance of complete classes of institutions.

These reservations being understood, we will mention a few more or less obvious cases in which degeneration does retrace the footsteps of progressive evolution.

1. *Tithings, hundreds and counties in England.*— In the chapter dealing with the pathway of degeneration in *Transformisme social*, G. Degreef mentions the following interesting facts :—

"Mr Herbert Spencer, after describing the formation of tithings, hundreds and counties in England under the Anglo-Saxon regime, observes that the tithings along with their courts of justice were the first to disappear, then the hundreds followed, though some vestiges of their old courts of justice remained, and only the counties and the county courts were left intact. Now we have historical proofs that English counties along with their courts of justice were created before the hundreds, and the hundreds before the tithings."[1]

2. *Order of elimination of various racial elements in a country.*—In his interesting work *Civilization et dépopulation*,[2] Dumont mentions certain facts which go to show that the inhabitants of poor districts, who are nevertheless of pure racial descent, have a birth-rate higher than that of the members of the population who are not aboriginal, and who

---

[1] Degreef, *Le transformisme social*, p. 450.
[2] P. 156.

for the most part dwell in the towns and fertile plains. From this he concludes that the various racial elements of a nation are eliminated in inverse order to that in which they were introduced. In France, for instance, the Frank has been completely absorbed in the Gaul.

3. *The degenerative evolution of political organizations.*—The progressive and degenerative evolution of political organizations has been described by Herbert Spencer as follows [1]:—

"Political integration, as it advances, obliterates the original divisions among the united parts. In the first there is the slow disappearance of those non-topographical divisions arising from relationship, as seen in separate gentes and tribes—gradual intermingling destroys them. In the second place, the smaller local societies united into a larger one, which at first retains their separate organizations, lose them by long cooperation; a common organization begins to ramify through them. And, in the third place, there simultaneously results a fading of their topographical bounds, and a replacing of them by the new administrative bounds of the common organization.

"Hence, naturally, results the converse truth that in the course of social dissolution the great groups separate first, and afterwards, if dissolution con-

[1] Herbert Spencer, "Political Institutions," Part iv. of *Principles of Sociology*, p. 286.

tinues, these separate into their component smaller groups. Instance the ancient empires successively formed in the East, the united kingdoms of which severally resumed their autonomies when the coercion of keeping them together ceased. Instance again the Carlovingian Empire which, first parting into its large divisions, became, in course of time, further disintegrated by subdivisions of these. And when, as in this last case, the process of dissolution goes very far, there is a return to something like the primitive condition, under which small predatory societies are engaged in continuous warfare with like small societies around them."

We may conclude then that political integration is attended by degeneration; primitive institutions disappear and make way for fresh institutions, and their disappearance is permanent. In the course of the dissolution of the Carlovingian Empire there was no reappearance either of the *gentes* or of the primitive tribal system; but when this vast organization broke down, it was natural that the more recently formed social bonds, having had the least opportunity of becoming consolidated, should be the first to be sundered.

4. *Degeneration in monetary systems.*—The principle that degeneration retraces the steps of progress applies equally to a very different range of ideas,—the evolution of monetary systems. Stanley Jevons says that there is little doubt that every system of coinage was originally identical with a

system of weights. A survival of this primitive condition existed in Roman law, and even when no use was made of them, the custom of bringing a pair of scales survived as a legal formality in the sale of slaves at Rome.

After the time of the Punic wars, the æs, which originally equalled a Roman pound in weight, diminished rapidly, until it became reduced to the weight of an ounce. The Romans had naturally reverted to weighing the metal, and the *æs grave* was money reckoned by weight, and not by tale. Generally speaking, whatever be the inconveniences of the method, currency by weight is yet the natural and necessary system to which people revert whenever the abrasion of coins, the intermixture of currencies, the downfall of a State, or other causes, destroy the public confidence in a more highly organized system.[1]

It is plain then that the more recent developments in the coinage system are the first to disappear.[2]

The disappearance of money altogether and the return to a system of exchange would represent a much farther advanced stage in degeneration.

[1] See Stanley Jevons in *Money*, International Scientific Series.
[2] There is no silver money and only a little copper in China. Nowadays, Mexican piastres, on reaching the country in payment of commercial transactions, are melted down into bars as soon as they fall into the hands of the merchants, and these bars are then imprinted with the Chinese stamp. This was the usual system employed amongst civilized peoples before the invention of money. See Thorold Rogers in "The Economic Interpretation of History."

5. *Degenerative adaptation in colonial legislation.*—In his treatise (*Annalisi della proprieta capitalista*), Loria furnishes another striking example of the law of degeneration: " When English colonies were first formed in America, the colonists hesitated to establish any legislation other than that of the mother-country. They were habituated to it; it was written in their own language, and therefore seemed best to correspond with their national characteristics. But, from the outset, the greatest difficulties were met with in the application of this legislation to the colonies.

" In the first place the Statute law of England, the most recent addition to the legislation, was found to be quite unsuited to the economic condition of a colony, and so common law alone came to be established, which, being the more ancient, was better suited to the social organization of a newly-formed society. But even this form of legislation did not remain permanent under social conditions profoundly different to those in which it had been originally established, and the construction of a special legislation was found to be necessary. In this way the common law of England came to be regarded as unsuited to her colonies, excepting in such cases as were unprovided for in the new colonial law."[1]

Loria then proceeds to give numerous examples of how these colonial statutes—owing to the simi-

---

[1] Loria, *Annalisi della proprieta capitalista*, ii. 48.

larity of circumstances between those for whom they were severally fashioned—came to resemble the primitive law of England.

6. *Degenerative evolution of the corporations of Western Flanders.*—In a treatise published by one of our number in May 1892, entitled *l'Evolution régressive des corporations de la West-Flandre*,[1] it was shown that among these associations, the institutions the first to degenerate were those most recently established.

The twelve corporations which still exist, though in more or less degenerate conditions in Bruges, Furnes, Eeghem, and Iseghem, were formerly constructed on similar principles and fulfilled the following functions :—

1. The furtherance of sociability (*i.e.* the holding of banquets and fêtes).
2. The encouragement of religious feeling (*i.e.* frequent celebrations of the mass and the building of new churches).
3. Mutual assistance (*i.e.* insurance against loss of work through illness, or against funeral expenses).
4. Mutual protection of professional interests.
5. The furtherance of certain political and military interests.

These various functions were established not simultaneously but in succession. The Flemish corporation of the fourteenth or fifteenth century

[1] *La Société nouvelle*, *Mai* 1892, Bruxelles, Monnom.

represents the last stage in a long series of corporations of different kinds, and of increasing complexity. It will be as well to glance briefly through these various kinds of corporations in order to compare the different stages of dissolution through which the corporative system eventually passed with the progressive evolution exhibited in the course of its establishment.

(*a*) Associations formed for the holding of banquets and fêtes (*convivia*) which were originally distributed throughout Northern and Western Europe.

(*b*) The *convivia* assumed a religious character when the Church, unable to suppress them, determined to transform them.

(*c*) Guilds (etymologically significant, according to Brentano, of repasts where all expenses were shared) where the original *convivium* was accompanied by religious ceremonies, but a feature of which was the addition of institutions of mutual insurance.

(*d*) Corporations (*ambachten*).—These embodied all the primitive institutions that had gone before: (1) the banquets; (2) religious ceremonies; (3) mutual assistance; (4) fresh means for the protection and development of professional interests.

(*e*) Finally, at the commencement of the fourteenth century, these corporations assumed both political and military functions. The concession of *keure* to the people in 1304 by Philip of Thiette was the first act by which was recognized the right of the corporations of Bruges to take part in communal administration and to provide a military contingent of their own.

When these corporations began to decline military and political functions were the first to be eliminated, in other respects the corporation continued to exist till the close of the eighteenth century in the form of economic groups, and groups for mutual aid and religious development.

An investigation of such corporations as still survive shows that it is in the economic functions formerly discharged that degeneration has made furthest strides, while the religious character is maintained and banquets are still held. Of the twelve corporations of Western Flanders six preserve the old corporative institutions almost intact; seven maintain a scale of charges and regulations connected with them; ten provide organizations for assurance against illness; eleven hold annual religious ceremonies; twelve, *i.e.* all of them, continue to hold banquets.

These figures alone show that the original and earlier functions have remained longest in force, but to further demonstrate this point it will be necessary to enter more fully into details and to study each group separately.

The first group—consisting of associations continuing to protect professional interests—consists of three branches: the four offices of Bruges (including vendors of lime, coal, seeds and beer), the two offices of Furnes (street porters and vendors of beer), and the community of the bakers of Bruges.

All the ancient statutes of the four offices have been preserved, including trade monopolies, the freedom of the city, and scales of charges, help in times of illness, accident, or when out of work; religious ceremonies and banquets. With the community of bakers, however, this is not the case; here the economic functions discharged are reduced to a minimum. The association continues to exist in a triple capacity: as a syndicate to keep up a fair price in wheat and bread, as a mutual assurance association in times of distress among the members, and as a confraternity imposing religious obligations and holding an annual banquet and fête.

In the second group consisting of mutual aid associations such as the corporations of tailors, shoemakers and weavers of Bruges, all trade interests have disappeared, and the corporations only exist in the capacities of confraternities and mutual aid societies. At this particular stage of degeneration these corporations resemble in a striking degree the old guilds which preceded the *ambachten en neringen*. It is interesting in connection with this, that the corporation of wool weavers (wollewevers) in Bruges has lost its orginal professional character, and quite heterogeneous elements have been introduced; there are only twenty-five weavers in Bruges, and their society numbers nearly two hundred members. The guild of weavers in Eeghem represents the third stage only—that of a religious confraternity. The office of mutual aid has disappeared, and only the Saints' Day Fête and the *convivium* remain : " The first Sunday after the fête of the Trois-Rois, which is the annual fête day of the weavers, the guild proceeds to the parish church, headed by a banner, accompanied by a jester, and with drums beating, where mass is celebrated in honour of Saint Severin. The rest of the day is passed in diversions."

The guilds of Saint Crispin at Iseghem, all that remains of the old corporation of shoemakers, are representative of the final stage of degeneration : the association has resumed its most primitive character, and is reduced to a mere dining society— the primitive *convivium*.

There are from 1500 to 2000 shoemakers in Iseghem. At the time of the Revolution their *ambacht*, having resumed its archaic form, was divided up into six or seven guilds. Some years ago a vain attempt was made to reconstruct these guilds and adapt them to modern requirements. At a time when success in this project was still hoped for, one of the promoters of the reconstruction wrote as follows :—

"The members of the guild still recognize Saint Crispin as their patron saint, but they no longer assemble in church to do him honour by the celebration of mass. They keep the anniversary instead by going from one public-house to another, with flying banner and drums beating, and the day has become merely a day of copious libations, and serves as a pretext for poor workmen, the fathers of families, to spend all the week drinking in public-houses."

In conclusion then the cycle is complete; a corporation, unless suddenly dispersed, ends as it began; in the last stage of decline it resembles the associations from which it originally developed, the most recently established functions having been the first to decay and disappear.

§ 2. *A criticism of the supposed inverse path of degeneration.*

These few examples suffice to show that in certain cases the more recently formed institutions are the first to decline and disappear, while the older persist to the end.

It must be remembered, however, that the contrary is at least as frequently the case. All changes of legislation, either juridical or religious, follow, but never precede, the economic transformations to which they relate, whether these be social or ethical, unless the transformations are ephemeral. " Imitation," says Tarde, " proceeds from the more obvious to the less obvious; that is to say, ends and feelings are imitated sooner than their means and expressions."

Title-deeds and armorial bearings survive nobility; houses continued to be held as personal or moveable property long after the disappearance of nomadic tribes, which, living as they did in tents, originated the conception.[1]

Among peoples where the system of marriage by groups has existed, family nomenclatures continued to persist long after the disappearance of the family system to which they owed their origin. "The family," says Morgan, "is an active element, never stationary; it keeps pace with the development of society in the march of progress. On the other hand, the reckoning of kinship changes very slowly; only after long lapses of time does it register the progress actually made by the family in the course of ages, and does not undergo any radical transformation until long after the family itself has been completely changed." "And," adds Karl Marx, whose critical annotations on Morgan's book were carefully preserved by Engels, "this also applies to systems of politics, law, religion, or philosophy."

These systems, formed after the completion of the social organization which they express, survive after the organization itself has disappeared. Their elimination is not of such importance to society as is that of the economic or family institutions themselves, as these, when they become useless and disadvantageous, are a drag on future development.

---

[1] Viollet, *Histoire du droit civil fr.*, p. 617.

"Although houses were for centuries treated as moveable property, they continued to be legally treated as such for a still longer period of time; it is characteristic of judicial ideas that they lag far behind economic progress."

It cannot be established, however, as a general principle, that the pathway of degeneration as regards societies or institutions, is inverse to that pursued by their progressive evolution. In the first place, the mere explanation of this supposed law shows that it is quite untenable.

What reason is given for supposing the decline of memory or will power, the degeneration of writing and speech, the decadence of societies and institutions, to be a retracing of the steps of progressive development? The reason given is that, other things being equal, the more fragile, unsteady and complicated structures are the first to fall.

Now, although the most fragile structures are frequently those most recently formed, and which have not had time to settle down and firmly establish themselves, it is also true that in many cases the more recent acquisitions and structures attain a more solid basis than those which have preceded them.

There is nothing invariable about the pathway of degeneration. It can no more be said to retrace the pathway of progress in an inverse direction than it could be said that in a country abandoned by its inhabitants the more recently formed paths of communication would be the first to become effaced. It is quite true that the broader roads, which would naturally last longer, are frequently the oldest paths of communication; whereas the footpaths, which are the first to disappear, are usually of more recent origin. Very frequently, however, the new roads follow a rather different direction, and, although more recently constructed, are not the first to disappear.

It is the same with great commercial crises. It is quite inexact to say with Ribot, who is responsible for the analogy: " Old houses offer the best resistance to the storm; it is the new houses which, being less solid, crumble and fall." [1]

After the time of the cotton famine, during the American war of Independence, the greater part of the old firms of Gand became bankrupt, whereas most of the large, newly-established joint-stock companies survived the crisis.

Moreover, in those cases where the most recently formed structures are the first to decay, it cannot be deduced that evolution is reversed, and that the institution returns to its primitive condition, for there is no reappearance of the intermediate structures.

[1] Ribot, *Les Maladies de la Memoire*, p. 99.

PART II

The irreversibility of degenerative evolution

MOST authorities on the subject are agreed that evolution is not reversible,[1] and that institutions or organs which have disappeared or been reduced to rudiments do not reappear and develop afresh. It would be a useless extension of this volume to cite many facts in favour of a view which is almost without supporters, but it will be useful to examine the exceptions, real or apparent, and to discuss—

1. If an institution or organ which has disappeared may reappear.
2. If an institution or organ which has been reduced may resume its primitive function.
3. If an institution or organ which has been reduced may redevelop and assume a function other than its original function.

[1] L. Dollo, *Les lois de l'Évolution* (*Soc. Belg. Géol. Paléont. Hydr.*, t. vii., 1893, procès-verbaux, pp. 164-166.

CHAPTER I

DO INSTITUTIONS OR ORGANS WHICH HAVE DISAPPEARED REAPPEAR?

Section I

Disappeared organs

In biology we are almost unaware of indisputable examples of the normal reappearance of disappeared organs.

1. *Plants.*—As the embryonic development of plants is usually direct, it is impossible to decide whether an organ which forms a component part of the embryological history represents an ancestral organ. However, in a few rare cases, artificial selection causes an actual reversion of evolution.

Typical geraniums possess two whorls of five stamens, as, for instance, in *Geranium*. In *Erodium* there is only one cycle of five. In *Pelargonium* one cycle of five is complete; the other is represented by two stamens and three filaments which have lost their anthers. But in certain varieties with very large flowers the two complete cycles reappear, five stamens having long, and five short, filaments. In this case there is no doubt as to the reappearance of the three stamens lost in typical *Pelargoniums*.

This reappearance is the result of artificial selection. The typical *Pelargoniums* have a bilateral symmetry, but horticulturists set a higher value on flowers with radial symmetry, and in consequence have produced flowers with such a symmetry. As a matter of fact, they have paid attention only to the symmetry of the petals, but in modifying that, they have also modified the symmetry of the stamens.[1]

The *Privets* (*Ligustrum*), like most of the Oleaceæ, possess only two stamens. Nevertheless, it is not uncommon to find among normal flowers of the common *Privet* specimens with three or four stamens. In this case, however, it is uncertain whether there is a real reappearance of lost organs, or if the loss has not actually become complete.

Animals.—In the case of animals, teratology and embryology furnish a few exceptional cases of an apparent reversion of degenerative evolution.

As an abnormality in the horse, the first, second, fourth, and fifth digits may reappear.

Adult man has lost the complete covering of downy hair. According to Ecker, however, hypertrichosis is an abnormality really due to the reappearance of this ancestral condition, as may be seen from the mode of its distribution in whorls. Hypertrichosis is a trait frequently inherited, and in this connection the Mauchamp variety of Merino

---

[1] The flowers upon which these observations were made were kindly provided by H. Cannell of Swanley.

sheep are interesting. This variety was obtained by breeding from a sport which appeared in a normal flock, and which transmitted its peculiarity to its descendants. From time to time, in normal flocks, variations occur which are similarly capable of giving rise to Mauchamp breeds.

As recorded by Willett and Walsham, there have been found in human children cases of a bone stretching from the scapula to the sixth and seventh cervical vertebrae. According to these authors, the bone represents the suprascapula of the tailless amphibia, which the normal homologue in man is the merest edge of the scapula ossified from a separate centre.

Such cases, as well as cases of polydactylism and of supernumerary mammæ, are usually set down as atavistic. However, the attempt to explain by atavism such pathological and teratological peculiarities must be made with caution. Such inherited anomalies occur very frequently in degenerate families—the neuropathic families of Féré[1]—and are associated with other abnormalities equally heritable and certainly not due to atavism. Such are pigmented retinitis, congenital cataract, chromatic asymmetry of the iris, asymmetry of the pupil, ichthyosis, pigmented erectile spots on the skin, and congenital disposition to bleeding.

As in a degenerate line of heredity these abnormalities may replace one another indiffer-

[1] Féré, *La Famille nécropathique*, 1894. Paris, F. Alcan.

ently it is very doubtful if any of them ought to be regarded as the inheritance of ancestral traits. Teratology contains no undoubtful case of the reversion of evolution.

On the other hand, it seems certain that in the individual development of some species there is a real reappearance of lost organs. The larval history of certain Crustacea Malacostraca[1] (cray-fish, shrimps, etc.), seems to provide instances.

In Stomatopoda, the youngest erychtheus larvæ (fig. 71, A) are formed of three parts: the head, five anterior thoracic segments, each bearing a pair of biramous swimming limbs (i.-v.), the three last decreasing in size from before backwards (these five pairs of appendages represent the five pairs of buccal appendages of the adult); three terminal posterior segments (vi.-viii.), and a caudal fin, all without appendages. In older larvæ the first and second pairs are profoundly modified, losing a joint and acquiring gills (fig. 71, B, i.-ii.); the third, fourth, and fifth pairs disappear completely, or at most occur as minute saccules (iii., iv., v.). New thoracic segments are formed, and later on in the third larval stage the thoracic appendages reappear in their final form.

Similar facts occur in the development of the Decapoda Macroura, such as *Palinurus* and *Scyllarus*. While within the egg the creature passes successively through nauplius and phyllo-

---

[1] Lang, *Anatomie comparée*, vol. i., p. 458.

some stages, and possesses all its thoracic appendages (three pairs of jaws and five pairs of ambulatory appendages). In this larval life the exopodites of the second and third pairs of jaws

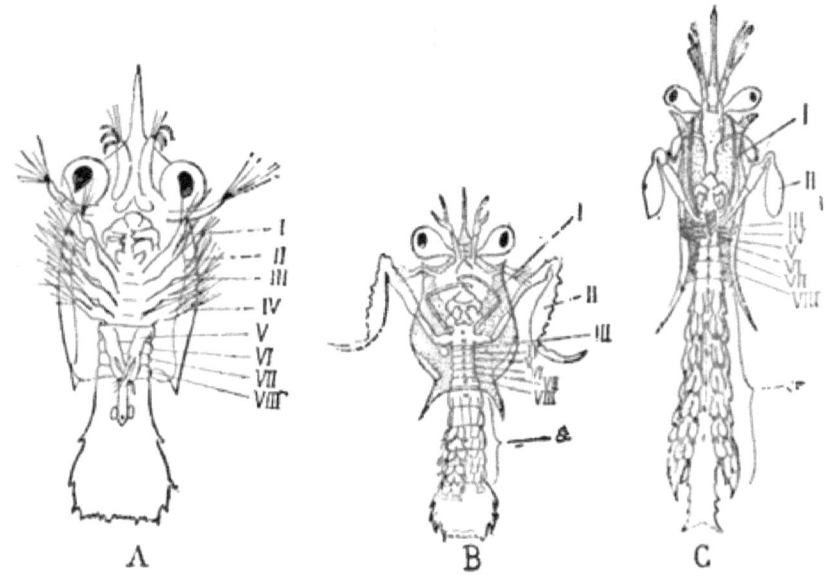

FIG. 71.—*Erychtheus larvæ of Stomatopoda.*
A, The youngest known *Erychtheus larva*: I—V, buccal appendages well developed.—B, Young *Erychtheus larva*: I and II modified buccal appendages; III—V, degenerating buccal appendages; VI—VIII, segments devoid of appendages.—C, Older *Erychtheus larva*: I—II, modified buccal appendages; III—V, reappearing buccal appendages; VI—VIII, ambulatory appendages in course of formation. (After Claus, in Lang's *Traité d'anatomie comparée*, vol. I.)

atrophy. Shortly before hatching, the first pair of jaws atrophies completely; the two pairs of antennæ and the two posterior pairs of ambulatory legs become very degenerate. In the young phyllosome stage these organs are rudimentary, but in the older phyllosome larvæ the first pair of jaws and the two posterior pairs of ambulatory limbs are reformed; the posterior two pairs of jaws reacquire exopodites, and gills are formed on the ambulatory limbs.

As a matter of fact, in most crustacea, the ambulatory appendages appear when they become necessary, that is to say at the end of larval life, but in the Stomatopoda and in the Decapoda Macroura, owing to inheritance, they appear much sooner. But when these appendages are useless during the larval life they disappear again to reappear at the end of larval life as in most crustacea. This adaptation of the larva to special conditions is of great importance, as the larval life is most important from the point of view of the species.

Section II.

Disappeared Institutions.

The apparent revival of bygone institutions.—It seems, at first sight, as if there were many instances of the subsequent revival of bygone institutions.

Those of ancient Rome and Greece, for instance, appear from time to time to have been reconstructed. In feudal Rome of the fourteenth century, Cola di Rienzi, by turns tribune and senator of the people, re-established the old republican constitution. During the Renaissance period the ancient schools of science were resuscitated, and during the French Revolution similar attempts at revival were made, especially in the department of politics. When Hérault de Séchelle, being ordered to draw up a scheme of legislation, revived the laws of Minos, in the constitution of the year VIII., the Tribunal, Senate and Consuls reappeared. During the First Empire, Napoleon, in imitation of Augustus, affected a respect for republican institutions, and had the coinage stamped with his own effigy and that of the Republic. In Germany, the Holy Empire which nominally ceased to exist in 1806, reappeared in 1871. In Greece, the Olympic games, suppressed in 1525, were re-established in 1896.

It is hardly necessary to insist upon the essentially superficial nature of these revivals. It is always possible to bestow upon new social systems the ceremonial code of an institution long since abandoned, but it is quite impossible that the institution itself should be resuscitated in the midst of surroundings which have been completely transformed. The consuls of the year VIII. and the emperors of modern times do not resemble the consuls and *imperatores* of ancient Rome more than the Christian societies of the present day resemble those of the middle ages. With regard to outer form in the drawing up of statutes, in all which constitutes, so to speak, the decorative side of the institution, the organizers imitated minutely the *keures* of the old *ambachten*; beneath this apparent similitude, however, were structures of an essentially different nature. Just as the new railway station at Bruges, in spite of its towers and its pointed arches, is far more like any other railway station than a Gothic cathedral, so the Christian societies of to-day, in spite of the archaic caprices of their founders, resemble more closely modern associations than ancient associations. We see then that it cannot be said in any of these cases that the actual revival of a decayed institution took place. The empty form reappeared, but the foundations and the essential parts had become completely transformed.

2. *The apparent disappearance of institutions.*—There are other instances, however, showing the opposite of this phenomenon. The essential parts remain unchanged, but the form itself is modified; the institution persists, but its existence is dissembled. Sometimes even when the dissolution of an institution has been enforced, there has been a reconstruction on the first opportunity.

An instance of this is the reappearance of polygamy among the Mormons, the last traces of polygamy having disappeared during the middle ages from the people of the West.[1]

[1] In Bigorre there was maintained up to the fifteenth century, a kind of system of concubinage called *massipia*, which, though not actually polygamy, was the contraction of an inferior union in conjunction with real marriage.

In Marseilles, too, polygamy seems to have reappeared in the middle ages, owing to the frequent intercommunication between

In ceasing to be legal, however, polygamy has by no means ceased to exist, even in the present day, and the Mormons in instituting polygamy have only given official recognition to what had never really disappeared.

In the process of social transformation, periods of transition are frequently characterized by reactions in favour of bygone institutions, which reappear although apparently permanently abolished. This has been the case with certain corporations, and with numerous other institutions of the old regime which reappeared after the revolutionary crisis. Some years ago a large landed proprietor in the domain of Peterhoff, in Russia, attempted, in the interests of his serfs, to introduce the rural system of European countries. He divided up the land into independent allotments, and built at his own expense a separate house for each family; but no sooner was the abolition of serfdom declared than the peasants proceeded to re-establish the primitive community, and to rebuild their houses on the old sites in spite of the considerable labour which this entailed.[1]

3. *Instances of convergence.*—It sometimes happens

that town and the East. It was never, however, established there officially as was the massipia in Bigorre. The municipality promptly suppressed it by issuing the following mandate: "Quod (vir) non habeat duas uxores, vel mulier duos viros." (Viollet, *Histoire du droit civil*, fr. p. 388.)

[1] De Laveleye, *De la Propriété et ses Formes primitives*. Paris, F. Alcan, 1882, p. 23.

that after the lapse of several centuries, an institution seems to reappear.

An exact analogy to the primitive *contubernium* (the community of the cabin) is exhibited in modern slavery. Only a few years ago, in the Spanish Antilles, marriage between slaves was recognized by neither church nor state. When a negro wished to become united to a particular negress he asked permission of his owner to share his cabin with her, and these unions could only be dissolved with the consent of the master. It is hardly likely that these slave marriages of the Spanish Antilles are survivals or rather resuscitations of the Roman *contubernium*. They rather represent a case of convergence: identity in circumstance has been productive of identity in institution. In this case, as in all others of the same kind, it cannot be said that a bygone institution has reappeared, for the new institution has quite a different origin. Further, in the other instances which have been mentioned, an institution which has reappeared has never really ceased to exist; a real dissolution has never been followed by a resuscitation. For this to happen, the whole social surroundings would have to be transformed into their former condition, which is obviously impossible.

CHAPTER II

CAN RUDIMENTARY INSTITUTIONS OR ORGANS RESUME THEIR PRIMITIVE FUNCTIONS?

THERE is no break of continuity between a rudimentary organ and the complete ancestral organ: the rudimentary organ is the ancestral organ transmitted by inheritance. It is the same organ not only because it has the same form, but because it is actually a part of it. For the reduced organ to return to its ancestral condition and resume its ancestral function, it must retrace the series of steps in degeneration along which it has passed.

So also in sociology there are many instances of institutions which are rudimentary, but which retain their original form because of uninterrupted imitative transmission. In their case also resumption of the primitive functional activity would imply a retracing of the degenerate steps.

This necessity of retracing shows at once that after a certain amount of degeneration resumption becomes impossible. We shall find in the few cases we are able to adduce, that when organs retrace their steps and resume their ancestral functions, degeneration had not gone very far in them.

SECTION I.

Rudimentary organs.

I. *Animals.*—Among animals it is very unusual for a rudimentary organ to become active again.

The only cases show that in them no great amount of degeneration had taken place.

1. *Muscles of the ear in man.*—It is known that the human ear possesses a number of intrinsic and extrinsic muscles reduced to delicate fibres, and incapable of producing movement of the whole ear or of one part of the ear on another part. In some abnormal persons, however, certain of these muscles may be well developed, making movements of the ear possible.

2. *The abdomen and appendages in deep-sea hermit-crabs.*—The hermit-crabs of the deep sea are another instance of reversion to an ancestral form. Littoral hermit-crabs inhabit the spiral shells of Gastropods, and to suit this mode of life the body is unsymmetrical, the appendages of one side being rudimentary. In the depths of the ocean such spiral shells are rare, and the crabs either abandon this mode of life or live in straighter shells. In consequence the limbs and the abdomen become nearly symmetrical again. It is plain, however, that in the littoral crabs these structures are not truly rudimentary.

II. *Plants.*—In plants it is very difficult to distinguish between the reappearance of lost organs and the formation of new organs.

1. *Hermaphrodite flowers in Melandryum.*—The Hermaphrodite flowers of *melandryum* (fig. 59) may be flowers which after being unisexual have again become hermaphrodite, or they may have retained

the primitive type. We cannot decide between the alternatives.

2. *Branches of Colletia cruciata, Crataegus, Vicia Faba, etc.*— Some cases, however, point clearly to a renewed development of rudimentary organs. Here are some examples. (See also, further on, page 244 on hybrid individuals of *Pentstemon*.)

Colletia cruciata (fig. 72) in the normal adult condition bears large flattened branches, which serve for assimilation and possess only very rudimentary leaves. Sometimes, however, the plant may give rise to more slender branches with normal assimilating leaves. These branches and leaves are probably the reappearance of the ancestral condition.

Wild pear and apple trees produce small lateral branches which are transformed into spines. These thorns have evidently arisen from normal lateral branches which originally bore leaves. In the cultivated varieties these lateral branches have resumed the leaf-bearing habit.

In the hawthorn (*Crataegus*) the lateral branches are similarly modified into spines. None the less, while these spines are still young they may be artificially stimulated to produce leaves by cutting the principal stem.

The branches of *Vicia faba* bear low down a set of rudimentary leaves. If the main stem be lopped while still quite young, the usually rudimentary leaves grow to the normal size.[1]

[1] Goebel, *Beiträge zur Morphologie und Physiologie des Blattes*. Bot. Zeit., 1880.

FIG. 72.—*Colletia cruciata* (after Goebel, *Pflanzenbiologische Schilderungen*, vol. I., p. 17). A, normal branch. B, branch recurring to ancestral form; (natural size).

Many plants such as *Sempervivum*[1] (fig. 73) normally possess a very much shortened stem. However, if the plant be grown in a saturated atmosphere the internodes of the stem lengthen out.

FIG. 73.—*Sempervivum tectorum*.
A (left figure), normal branch; B (right figure), branch grown for several months in a saturated atmosphere.

Certain *Veronicas* (fig. 74) bear only small scaly leaves. Cultivation of these plants in an atmosphere saturated with water, results in the appearance of normal leaves.

Lastly, we may quote again the instances given by Goebel of the production by *Equisetum arvense* of normal leaves under special conditions.[2]

[1] The branch figured here was grown by G. Clautriau, in the Brussels Botanical Institute.
[2] Goebel, *Ueber die Fruchtsprosse der Equiseten.* (*Ber. d. d. Bot. Ges.*), vol. iv. p. 184, 1886.

Section II.

Rudimentary institutions.

It seems, at first sight, as if some societies of the present day furnished instances of a return to primitive conditions.

Modern developments in finance seemed to tend towards a return to the exchange system. Political institutions, after a period of absolutism, point anew towards democratic equality. Corporations reappear in the form of syndicates or religious societies. Landed property, formerly collective and now individual, seems to be again tending towards collectivism.

Fig. 74.—*Veronica cupressoides* (after Goebel, *Pflanzenbiologische Schilderungen*, vol. i. p. 19). End of a branch grown under a bell-jar in a saturated atmosphere. In the older parts (B), the leaves are small and applied to the stem; in the younger parts (A), the leaves are larger and protrude from the stem.

The same phenomenon occurs in the evolution of matters relating to maritime rights. The sea, according to Roman law, was equally open to all maritime nations. Later on it has been from time to time practically in the hands of a few nations, and we have now returned to a condition in which it is equally open to all.[1]

[1] Tarde, *Transformation du droit*, pp. 161-162. Paris, F. Alcan, 1890.

These, however, are not cases of true revival. The resemblance goes little farther than the name.

A comparison of modern institutions with such survivals of primitive institutions as continue to exist will demonstrate this point. The difference between them is so wide that it would be hardly possible to utilize the old as a basis upon which to form the new.[1]

---

[1] Cf. Durckheim, *Les Règles de la méthode sociologique.* Paris, F. Alcan, 1895.

In sociology, dealing as it does with things familiar to us all, such as the family, property, crime, etc., it is useless to attempt to adhere to exact definitions. The exact meaning of some words in common use in conversation cannot be defined with any precision; the common acceptation of these words cannot be avoided. Now this common acceptation is frequently very ambiguous, so that two totally different things are often referred to under the same name, causing hopeless confusion.

There are, for instance, two different kinds of monogamous unions—those so only in point of fact, and those which are legally so. In the first case, a man has only one wife, though legally entitled to several; in the second he is only legally entitled to one. These two kinds of conjugal conditions are quite different, and yet the same word serves to express both; it is commonly said of some animals that they are monogamous, although there can be nothing approaching to a legal contract between them. Spencer, when dealing with the subject of marriage, makes use of the term monogamy without defining it in its common and equivocal sense. The result of this is that the evolution of marriage seems to him to represent an incomprehensible anomaly. It seems, according to him, that the superior or monogamous form of union was prevalent during the primitive phases of historic development; that it then disappeared during an intermediate period, to subsequently reappear. From this he concludes that there is no regular connection between

It may be definitely asserted then that a reduced, but still persistent, institution never again becomes actively functional. The following are a few examples which will serve to illustrate this point:—

1. *The truck system and clearing-house.*—Some forms of the primitive system of exchange survive, not only in countries where money is unknown, but in certain industries where the workers continue to be paid in kind (the truck system).

On the other hand, there seems to be a modern tendency towards the elimination of money as an instrument of exchange. The clearing-house system is singularly analogous to the old exchange system. "The truck system," says Stanley Jevons, "represents the first and the last stage; but it appears for the second time in a very different form. Gold and silver money continue theoretically to be the instrument for buying and selling, but practically metal no longer constitutes the real medium of exchange, and has ceased to pass from the hands of the purchaser into those of the vendor."

In this transformation there is obviously no return to primitive systems, the last vestiges of which, far from being revived, are rapidly disappearing.

social development in general and a progressive advance towards an improved system of family life. A more exact definition would have prevented this erroneous conclusion.

The suppression of the truck system coincides with the development of the clearing-house system.

2. *Corporations and syndicates.*—The radical differences existing between the corporations of former days and the greater part of modern professional associations has already been pointed out. The ecclesiastical associations, however, of the present day are modelled as closely as possible upon mediæval institutions. It does not follow that the last remaining vestiges of the latter have been revived. There seems to be evidence that quite the contrary has taken place.

At Bruges no attempt was made by the founders of the guild of *ambachten* to resuscitate such mediæval corporations as continue to exist in a state of decline.

At Iseghem, a small town in the west of Flanders, we have already seen that the corporation of shoemakers was divided up into six or seven guilds at the time of the Revolution. An attempt was made a few years ago to reconstruct and modernize these guilds, but the scheme fell through. A new corporation—wholly disconnected with the guilds of Saint Crispin, and with no structural resemblance to them—was established instead.[1]

3. *Archaic collectivism and modern collectivism.*—Societies of the present day exhibit numerous

[1] Emile Vandervelde, *Enquête sur les Associations professionelles d'ouvriers et d'artisans en Belgique*, i., p. 17, Bruxelles, 1891.

vestiges of archaic collectivism. The question arises as to whether there is a tendency in the modern school of collectivism to resuscitate such vestiges as remain of the old archaic form of collectivism. Far from this being the case, collective property, as conceived by the modern socialist, implies the suppression of the few existing remnants of archaic collectivism.

Inheritance, *ab intestat*, for instance, is a survival from the days of the family community, which itself arose, as we have already seen, from the primitive community. If the modern collectivist school had any desire for a return to the old primitive community, it would make for the reconstruction of the family community by re-establishing the law of collateral succession. Now it is just the opposite with the collectivists. In order to establish a universal system of collective property they demand among other things, the suppression of inherited succession, *ab intestat*, at least as regards the collateral line of descent.

4. *The survival of elective sovereignty in England.*—The above examples apply to institutions which have degenerated without having completely ceased to be functional.

It very rarely happens, however, that having arrived at that condition, they renew their vitality and all their former functions, and this still more rarely occurs in cases of genuine survival.

In the English coronation ceremony vestiges

remain of the old democratic system in which the king was elected by the people.[1]

The English sovereignty of the present day is merely a decorative institution, the real head of the Government being the Prime Minister, who is nominated in fact if not in theory by the public. This system may almost be regarded as a return to bygone democracy. Nobody would wish, however, to revive the old system of elective sovereignty, and to retrace in an inverse direction the various stages of its degeneration.

CHAPTER III

CAN RUDIMENTARY ORGANS OR INSTITUTIONS REDEVELOP AND ASSUME NEW FUNCTIONS?

THE few facts which we are able to cite on this subject must be received with considerable caution.

[1] The formality of an election disappeared during the Tudor period. The coronation of Henry VIII. was the last occasion on which the formula was read which set forth the national agreement with and recognition of, the succession. The king was, in fact, declared chosen and elected. This formula of election, which disappeared after the coronation of Henry VIII., is recalled to mind by the conclusion of the coronation ceremony of the present day. The archbishop, walking in succession to all four corners of the platform upon which the throne is placed, addresses the people in the following terms: "Gentlemen, I herewith present to you the undisputed sovereign of the realm. Come all who are present and offer homage to him. Are you prepared to offer it?" and the people signifying their assent by acclamation, cry, "God save the Queen" or "God save the King." (De Franqueville, *Le gouvernement et le parlement Britanniques*, i., p. 291.)

Section I.

Rudimentary organs.

1. *Animals.*—In *Birgus latro* (a land-crab of the Philippines), the gills are atrophied and the bronchial chamber is very richly supplied with blood vessels, while a kind of incipient lung is formed from the lining membrane of the reduced bronchial chamber.[1]

However, it is by no means certain that the atrophy of the bronchial apparatus has preceded this development of a pulmonary apparatus. In the following case it rather seems to be one in which a rudimentary structure has redeveloped in order to assume a new function. In the development of the urinary organs, it appears that the ducts of the mesonephros are quite independent of those of the pronephros, although these mesonephric ducts become functional later in the embryonic life than the pronephric ducts. They are, nevertheless, formed at an earlier stage, and their rudiments have appeared before there is any trace of the others. From this fact it would appear that in some ancestors of existing vertebrates there existed simultaneously mesonephric canaliculi and canals homologous with them, but exercising a different function. Such a condition actually exists in *Amphioxus*: in the branchial region of that animal there are pronephric urinary canals and genital chambers which

[1] Semper, "The Natural Conditions of Existence as they affect Animal Life." (International Scientific Series.)

are homologous with mesonephric spaces; but these latter do not exist as genital chambers in higher vertebrates.[1]

It must be noted that the homology between mesonephric spaces and the genital spaces of *Amphioxus*, as made by these writers, is not universally accepted.

2. *Plants.* — The Scophulariaceæ, which have usually four stamens, are derived from ancestors which possessed five. Usually the fifth stamen is only represented by a tiny process which rapidly atrophies. However, in *Pentstemon* the fifth or posterior stamen is developed, not as a functional stamen, but as a staminode, the function of which is to stretch open the flower to make it accessible to hymenopterous insects with short proboscess.

Can it be said that in such cases a rudimentary organ has really become redeveloped to assume a new function? To establish this it would be necessary to show in the case we have just mentioned, that the stamen did not become transformed directly into a staminode, but that it first became rudimentary and then developed afresh into a staminode.

An interesting fact is, that in some hybrid varieties of *Pentstemon* the staminode becomes fertile

---

[1] Boveri, *Die Nierencanalchen der Amphioxus.* (*Zool. Jahrbuch. Abth. Anat. und Ontogenie der Thiere*, vol. v., 1892.)

Wiedersheim, *Grundzüge der Vergleichenden Anatomie der Wirbelthiere.* Jena, 1893.

again. In some flowers sent to us by Mr Cannell of Swanley, the number of petals was increased to six, seven, eight, or nine. In some of these the posterior stamen was sterile and like a staminode; in others there were five fertile stamens. It is obvious that in this case the staminode had resumed its original function after having lost it.

Section II.

Rudimentary institutions.

The Levirat.—In his work entitled *Tableau des origines et de l'évolution de la famille et de la propriété*, Kovalewsky mentions an instance of a reduced institution which, without having first ceased to be functional, became transformed into another institution.

"The custom mentioned in the Bible of alloting a woman to the brother of her deceased husband, is explained by the primitive condition of things with regard to the relations between the sexes; all the women were the common property of the men belonging to one group of relations. Under the name of *levirat*, this custom survived for several centuries, owing to the idea which arose later on that a wife was property. Consequently, on the death of the husband, the widow, along with his other belongings, was treated as the inheritance of the person whom the death promoted to the rank of chief or head of the family community."

The *levirat*, a family institution, thus derived from the old system of marriage by groups, was transformed by degrees into an economic institution. It is important to notice, however, that this transformation was effected without the institution having even been reduced to the condition of a mere survival.

PART III

SUMMARY AND CONCLUSIONS

From all the facts that we have brought together, the general conclusion becomes plain that retrogression, notwithstanding the etymology of the word, does not imply a return to the ancestral condition.

Rudimentary organs and institutions resemble the primitive states of these, in so far as they no longer possess certain parts which the primitive stages did not yet possess. None the less, profound differences exist between the primitive and the reduced forms. In the primitive condition the institution or organ is capable of varying in the direction of new uses; in the reduced form, after a certain degree of atrophy, there is no longer the possibility of redevelopment to resume old or to acquire new functions. These observations apply equally to biology and to sociology.

Magnan and Legrain, in their work on degenerate persons, came to similar conclusions. They came to regard degenerate persons as abnormal, chiefly because they were devoid of the power to reacquire the normal condition and quite unlike their primitive ancestors, who, although possibly brutal and unintelligent, were normal beings with the activity and stamina necessary for future progress.

The following two diagrams, borrowed from these authors, represent clearly the differences between the initial and reduced condition of an organ or institution :—

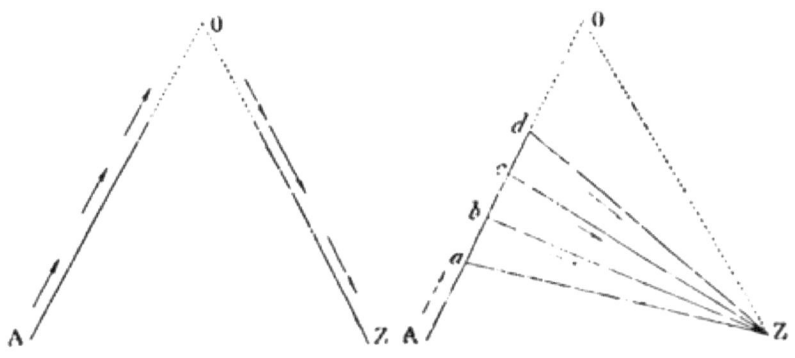

In the diagrams the ascending lines represent the progressive evolution of an organ or institution; the descending lines represent the degenerative evolution. From the point a, representing the primitive condition progressive evolution passes towards o, an imaginary perfect condition of the organ. Along the upward line, however, the points a, b, c, d, etc., represent obstacles to further progress—that is to say, factors tending towards degeneration. From these points lines of degeneration pass towards z, and the condition at z, although representing that at a, is not identical with a, and is not reached by a sliding backwards down the line o, a.

Thus, although the most recently acquired features may disappear first, degeneration is not an actual retracing of steps until the point of departure is reached. The degenerate condition is a new point, and really the term retrogressive evolution is misleading.

BOOK III

CAUSES OF DEGENERATIVE EVOLUTION

PART I

ATROPHY OF ORGANS AND INSTITUTIONS

The factors of atrophy

THE causes which are active in producing degeneration are various, but they may all be referred to the limited nature of the means of subsistence, that is to say, of nourishment in the case of organisms, and of capital and labour in the case of institutions. This limitation produces a struggle between the individuals (societies or organisms) and between their component parts.

In the course of the perpetual struggle for existence among the different parts of an individual, the institutions or organs which have ceased to be functional tend to disappear, their nourishment being absorbed by the active parts.

1. *Biology.*—In biology the struggle for existence among component parts appears clearly as a factor of degeneration in the case of accidental atrophy. This is to be seen, for instance, in the atrophy of

the leaves of an etiolated plant[1] or of the muscles of a limb which has been immobile for long, or in the case of muscles which have become inactive from disease of the central nervous system. The results are similar in cases of normal atrophy. In frogs, toads, and other Batrachia Anura,[2] the disappearance of the tail before the adult state is reached is the result of a struggle amongst the cells. The active protoplasm of the muscular fibres develops specially, and gives rise to many cells, which enter the contractile material and separate its elements. Gradually all the contractile material is absorbed by these isolated cells.

Many plants, especially *Sempervivum* (see fig. 73, A, p. 236) possess a reduced stem with the leaves closely massed upon it. This reduction of the stem, which is nearly constant in the

---

[1] When a cutting from a potato or a seed (fig. 75) is allowed to sprout in the dark, the young stems assume characters different from those of plants grown in light. The absence of chlorophyll produces important modifications of growth. In light the stem is short, and the leaves are large and expanded; in darkness the stem is very long, and the leaves are much reduced. This atrophy of the leaves is the result of the struggle for existence amongst the organs of the plant. Light increases the rate of transpiration, which is chiefly due to the presence of chlorophyll. As chlorophyll is most abundant in the leaves, the transpiratory current sets strongly towards them, carrying in it the nutritive materials for the formation of new cells. On the other hand, in etiolated plants, transpiration is slower, and the nutritive materials delayed in the stem give that the opportunity for specially active growth, which takes place at the expense of the leaves.

[2] Metchnikoff, *Annales Inst. Pasteur*, January 1892.

species, is the result of a struggle amongst the organs produced by a scanty water supply. The leaves attract to themselves the greater part of the water absorbed by the roots, and thus retard the growth of the stem. But if the plant be cultivated in an atmosphere saturated with water, the struggle between the leaves and stem is stopped, and the stem grows to a much greater extent (fig. 73, B, p. 236).

Vicia Faba, like most vetches, produces at germination rudimentary leaves, and similar leaves are borne at the base of each branch. Goebel has shown that these rudimentary leaves may be made to grow by cutting away the terminal buds at a young stage. In this way the struggle between the first formed and later leaves is suppressed.

FIG. 75.—Two seedlings of *Cicer arietinum*. The seeds were planted at the same time. A, seedling grown in light. B, seedling grown in darkness.

In the case of the individual, atrophy results from the struggle for existence amongst the organs. In atrophy throughout a species it is the struggle for existence amongst individuals that plays the chief part. Clearly in this struggle, useless organs become impediments and burdens. If any organs are useless they are harmful as they use nutrition without conferring any advantage upon the whole organism. Darwin pointed out that when through changed environment a structure became useless, its degeneration became certain, as it was a disadvantage to the individual to squander nourishment upon a useless part. Weismann has shown that such a reduction or disappearance of a useless organ is the result of variation and natural selection.[1]

Variation results in the appearance of individuals with the useless organ in various stages of imperfect development; natural selection perpetuates these advantageous stages by giving advantage to individuals which tend to produce the organ in the most degenerate condition.[2]

---

[1] See, however, Herbert Spencer's *Social and Moral Problems*, as he differs from Weismann on this point.

[2] The following are good examples of the operation of variation and selection in producing atrophy in a species:—

(1) *Loss of constant colour among domesticated animals.*—Wild animals, especially birds and mammals, have a colour which is constant for a whole species. Frequently the colour is protective in rendering the animal little distinguishable from the environment in which it lives. As soon as such a species has been domesticated man becomes its protector, and protective colouration is no longer necessary, and soon disappears. In the wild state, the colour is quite as variable as in the domesticated state, but the abnormal individuals become the prey of enemies and are removed from the species. This applies not only to animals which are preyed on by others, but to predatory animals themselves. The wild-cat, for instance, will have less difficulty in stalking its prey if its colour makes concealment easy. In the

2. *Sociology.*—With societies this elimination of useless structures is effected much more easily than with organisms for several reasons:—

In biology a special factor, heredity, gives to specific characters a force which does not exist in the same degree with social institutions. Now functional organs common to a whole line of descent are not easily effected by the influence of individual surroundings. Further, the trans-domesticated condition man provides food and the colour being unimportant all variations may survive.

(2) *Loss of spines in plants on oceanic islands.*—It is well known that the presence of spines protects plants from the ravages of herbivorous animals, particularly mammals. But in oceanic islands bats are generally the only mammalian inhabitants, and so, according to Wallace (*Darwinism*), there are no spiny plants in the indigenous flora of St Helena. The much richer flora of the Hawaian islands includes only a very few prickly plants. All the endemic genera are unarmed, as also are most of the endemic species of other genera; even genera like *Xanthoxylum*, *Acacia*, *Xylosoma*, *Lycium*, and *Solanum*, which are so frequently armed in other countries, are there represented by unarmed species. The two species of *Rubus* bear prickles reduced to the merest points and the two palms are devoid of spines. How is the absence of spines to be explained in these plants? The plants have been derived from the mainland, the seeds being brought by the wind, by currents, or by birds, and having found soil have germinated. In their new country they are not attacked by herbivorous animals, and it is immaterial to them whether or no they bear spines. The individuals badly armed are at no disadvantage compared with those possessing the normal armature; on the other hand, they have the advantage of being without useless organs to support. Spines, in consequence, gradually disappear.

This struggle for existence may cause the disappearance of some organisms themselves, and not only the atrophy of parts of them.

mission of acquired characters is, to say the least, doubtful. There is no proof of individual atrophy being hereditary, while with societies modification may be transmitted by imitation. Institutions which have fallen into disuse rarely recur in freshly-formed societies. Natural selection plays an all-important part in biology, but it is artificial selection which almost exclusively governs the social domain.

Many vegetables, as for instance the carrot (*Daucus Carota*) are natives of France. The seeds of the cultivated carrot must frequently be carried to waste lands or uncultivated soil. The domesticated variety, however, is never found wild although the wild variety is abundant. This vegetable has lost the power of struggling against weeds; it flourishes only when it is protected by man and when by repeated weedings its wild competitors are removed. When it is returned to its original wild haunts the plant dies out at once.

Most cereals, although we may not know their wild ancestors, are in a similar condition. For instance, if man were to cease cultivating Wheat (*Triticum sativum*), or Rye (*Secale cereale*), there is no doubt but that these would completely disappear. Their fate would be shared in Belgium at least by many species which are reaped with them at harvest, such as *Centaurea cyanus* the Corn-flower, *Agrostemma Githago* the Rose-Campion, *Specularia speculum*, and others. If a corn-field were left to the free operation of nature, weeds would soon intrude and cause the disappearance of the plants usually present in it. What would happen in Belgium would happen with other plants in other countries. Thus, near Bergen in Norway, some plants, such as *Melandryum album*, *Silene inflata*, *Vicia cracca*, etc., occur only in cultivated fields. In Java, many aquatic plants such as *Limnocharis Plumieri*, *Ludwigia perennis*, *Jussiaea suffruticosa*, etc., live only in the rice-fields which are artificially watered and manured. The cessation of tillage would cause the disappearance of all these plants from those localities.

An institution which has become useless and burdensome is generally suppressed before its complete degenerative evolution is accomplished. This suppression may be either voluntary—as in the liquidation of a commercial company, for instance—or it may be enforced. By the terms of Article 73 of the Belgian Company's Act, "the dissolution must be declared upon the demand of all those interested at the termination of six months from the time when the number of shareholders has been reduced to less than seven."

The downfall is generally effected in this sudden way, either voluntarily by the interested parties, or by the intervention of legislative means.[1]

Sometimes, however, artificial selection does occur,

---

[1] The occurrence of autotomy or self-mutilation in animals, as in crabs, has analogies with what we have been discussing. Similarly some plants brought into a new locality suddenly shed their leaves. *Ranunculus aquatilis*, cultivated in water, produces long divided leaves without stomata and with chlorophyll in the epidermic cells. If, from some chance, the water falls below the level of the plants, the adult leaves become dry and perish. The very young leaves growing unsubmerged are still divided, but to a lesser extent; they have stomata, and the epidermis is devoid of chlorophyll. If the plants be again submerged, this form of leaf dies, and there is a new development from the youngest leaves of the normal aquatic type.

Other plants show similar occurrences. Thus, when a *Fuchsia* that has been cultivated in a conservatory is exposed to the air, all its leaves are shed and are replaced by new leaves. These new leaves again fall at once if the plant be brought back into the conservatory. This is a real case of autotomy in plants.

and then the degenerative evolution of useless institutions is brought about in a similar way as that of non-functional organs. Atrophy of this kind may be, as in biology, either accidental or normal.[1]

In a besieged town cut off from all outside communication, all train service is necessarily stopped and the railway staff rendered useless. On the other hand, the defence of the city requires both men and money. Under these circumstances the resources of the railway naturally come to be absorbed in the service of the defence.

As a good example of normal atrophy may be mentioned the disappearance, at a certain point

---

[1] See Durckheim on the difference between normal and accidental sociological phenomena in *Les Règles de la méthode sociologique*.

"All sociological phenomena, like biological phenomena, are liable, while remaining the same individuals, to revert to different forms. Now, of these forms, there are two kinds of reversion :—

"The one is common to the whole species, and is to be found, if not in each individual, at least in the greater part of them. The cases are not always identical, varying slightly with the individual, but individual variation is restricted to very narrow limits.

"The other kind of reversion is exceptional, being of a nature rarely met with, and, when occurring, is seldom permanent throughout the life of the individual. Cases of this kind are exceptional in point of duration as in other respects.

"Here then are two distinct varieties of phenomena which should be distinguished from one another by different terms. Individuals exhibiting only common characters are called 'normal,' while those exhibiting exceptional characters are designated as 'morbid' or 'pathological.'"

of social development, of the popular assemblies in which lay the origin of future societies, *i.e.* the *comitia*, assemblies at the market-place, the *witenagemot*, May Day games, rustic assemblies, the *Landsgemeinde* of the Swiss cantons, the parochial assemblies of the Andora Valley, and the town meetings of New England, etc. Some of these were suppressed, but some of them merely fell into disuse as other institutions arose which were better adapted to more modern conditions of society. This happened with regard to the *comitia curiata* of the Romans during the period of the Empire, which were gradually supplanted by the *comitia centuriata* and tribunal comitia. At the time when they are first mentioned in history, they fulfilled only one function: that of ridding the laws of all traces of extrinsic customs.[1]

To sum up then, it is plain that although social degeneration is brought about by the same general causes as organic degeneration, the comparative importance of the factors in degenerative evolution is far from being identical in the two cases. The autotomy of organs, a protective self-mutilation, exhibits only a far-fetched analogy with the conscious and voluntary suppression of social structures which have become either useless or prejudicial.

Direct individual adaptation, which plays a part

[1] Mommsen, *Droit public romain*. *Le peuple et le Sénat*, vol. i., p. 364.

of only secondary importance in the development of animals and plants, is a dominating element in sociology. Institutions are able to modify their structure by assimilating new inventions and improvements, and by getting rid of the useless parts. On the other hand, indirect adaptation spread over a species plays no part whatever in sociology, for societies seldom reproduce the structures of the societies from which they sprang, when the latter have ceased to exist; whereas, in animals, when a useless organ is reproduced by hereditary repetition, variability and natural selection become agents in its suppression.

CHAPTER I

ATROPHY OF ORGANS

Part II

Causes producing atrophy

THE ultimate cause of the atrophy of organs is the limitation in the quantity of nourishment. We have shown that if there were an indefinite supply of food there would be no struggle, and, as a result, no degeneration. We have now to examine more minutely the course of atrophy and the nature of its immediate causes.

The atrophy of an organ is a reduction in

structure, in nutrition, and in functional activity, but the succession of these three events varies with the nature of the exciting cause.

Reduction begins with structure when the exciting cause is lack of space, due, for instance, to increase in another organ (*atrophy from lack of space*).

Atrophy begins with function when an organ has become useless (*atrophy from lack of utility*). This uselessness may arise from two causes; the function may be no longer useful to the individual or to the species, or it may be assumed by another organ.

Lastly, atrophy may begin with a diminution in the supply of nutritive materials (*atrophy from lack of nutrition*). This defective nutrition may be the result of a general cause such as feebleness of the whole organism, or it may be due to the hypertrophy of another organ.

§ 1. *Atrophy from lack of space.*

Cases of this kind are rare.

1. *Development of the teeth.*—Among animals, the development of the teeth furnishes an excellent example. The number of the teeth in human beings is reduced compared with the number present in Lemurs and in Platyrhine Monkeys. These have six grinding teeth while in man five is the maximum number. Our posterior molar, however, appears late in life; it is smaller than the others, so that it may be useless for chewing;

frequently it is absent, in the lower races in nineteen per cent. of cases examined; in the higher races in forty-two per cent.

The reduction in number and size of these teeth is due to a reduction in size of the jaw-bones, a cause which also frequently produces a distortion in the arrangement of the other teeth. The rudiments of the wisdom teeth appear on the maxillary tuberosity and on the coronoid process; it is only after eruption that they come into normal connection with the jaw-bones.

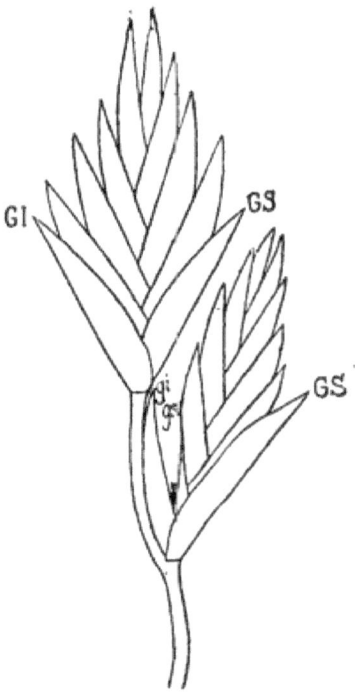

Fig. 76.—Tip of an ear of *Lolium perenne*, with two earlets, the superior bearing two glumes, the lateral earlet with only the superior glume. In the lateral earlet the lower flower is open and has two glumules; all the other flowers are closed and exhibit only the inferior glumule.
GI, inferior glume; GS, superior glume; *gi*, inferior glumule; *gs*, superior glumule.

2. *Atrophy of the superior glume.*—Among plants there are few cases of atrophy as a result of lack of space. In grasses of the genus *Lolium*, the earlets are arranged in a spike, but in such a fashion that only the terminal earlet has space for both glumes (fig. 76). The

superior glumes remain; the inferior, pressed against the axis, disappear after the embryonic development of the flower.

3. *Degeneration of paleæ and of stamens.*—Lack of space is probably the immediate cause of the disappearance of paleæ in the receptacles of some composite flowers and of the posterior stamen in the flowers of some Scrophulariaceæ and Labiates.

In normal racemose inflorescences each floret grows in the axil of a reduced leaf called a bract. When the axis of the inflorescence is shortened and the florets crowded, as in the capitula of composite flowers, it frequently happens that the bracts of the florets (termed paleæ) disappear. This absence is most usual when the capitulum is small and the florets are large.

In Labiates and most Scrophulariaceæ, although the ancestral number of stamens was five, there are not more than four present; when only one is absent, it is the original posterior stamen which was pressed against the axis of the inflorescence.

§ 2. *Atrophy from lack of use.*

1. Functional Inutility.

(1) *Etiolated plants and immobile limbs.*—We have already quoted as instances of accidental atrophy, cases of degeneration of leaves in etio-

lated plants, and of muscles in unused limbs (see fig. 75, p. 253).

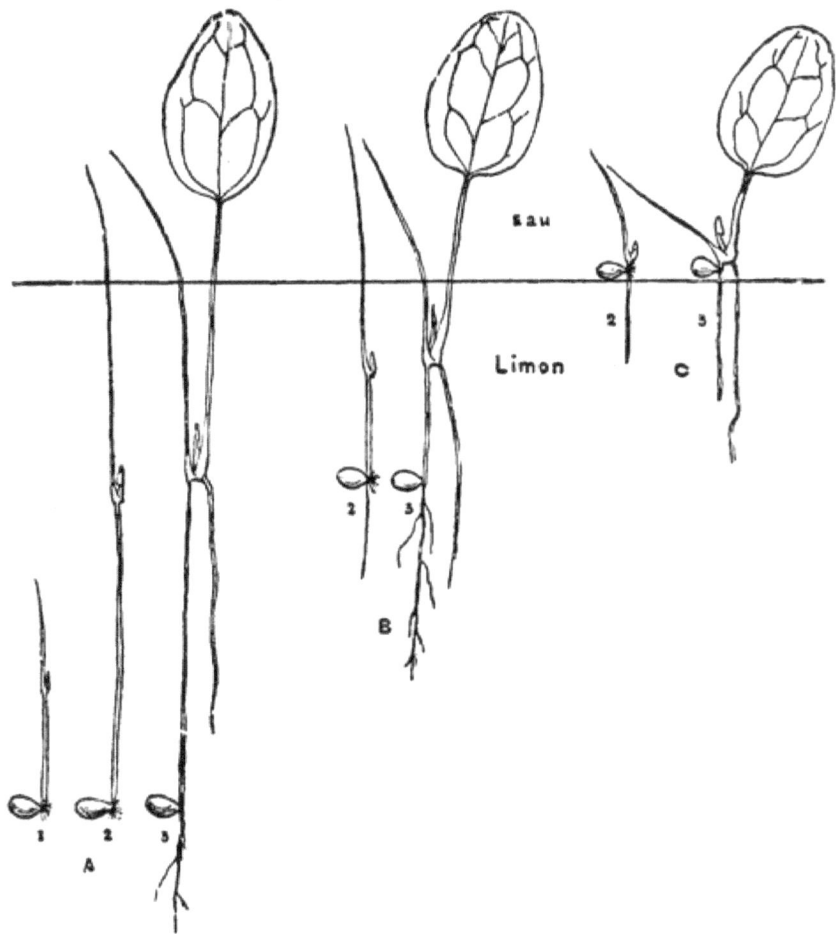

FIG. 77.—*Nymphaea alba* sown on the mud and at different depths in it. 1, 2, 3, successive stages of the same seedling. Eau, water; limon, mud.

(2) *Epicotyl and primary leaf of Nymphaea.*—These structures in the water-lily are good examples of normal atrophy in individuals. During germination (fig. 77) the cotyledons of the water-lily remain inside the seed, and a new organ (at right angles to them) grows vertically upwards. The lower part of this is the first internode of the stem (epicotyl), and the upper part is a primary acicular leaf. It grows upwards through the mud until the summit of the leaf reaches light. The growth of the epicotyl is then much slower, and its terminal bud begins to shoot out horizontally. The use of this growth of the epicotyl and primary leaf is to carry the bud to the light. When that purpose is achieved, these structures atrophy. A similar occurrence may be found in *Sagittaria sagittifolia* (fig. 40, H, I, p. 72). In that case, however, it is the hypocotyl which elongates, until light is reached, and then degenerates.

(3) *Roots of Utricularia, cotyledons of parasitic plants, leaves transformed to spines in Phyllocactus crenatus.*—As instances of atrophy throughout a species produced by inutility of the parts concerned, we have already mentioned the roots of *Utricularia* and the cotyledons of the parasitic plants *Cuscuta, Orobanche*, etc.

The spines of *Phyllocactus crenatus* produced from modified leaves are another example (fig. 78). Above the rounded base by which a branch of *Phyllocactus* is attached to older branches, there is an angular region, the sides of which are prominent, and bear leaves modified into spines to serve as protecting organs, as in the similar case of *Cereus*. Higher up the branch the prominent sides become flat, and the spines are replaced by minute scales. This degeneration is the result of loss of utility. The *Phyllocacti* are epiphytes, and their situation consequently places them out of the reach of cropping animals. The spines near the basis of the branches are a survival from the terrestrial ancestors of *Phyllocactus*. The spines higher up have degenerated.

Fig. 78.—Branch of *Phyllocactus crenatus*.

Animals offer many instances of atrophy as a result of inutility, both in individuals and in species.

(4) *Atrophy of the branchial arches in mammals.*—As they are no longer functional, most of the mammalian branchial arches atrophy. Three pairs alone persist, and of these it is only those parts which are useful.

(5) *Atrophy of ventral fins.*—Instances of atrophy through uselessness in species are to be found in the ventral fins of fishes like the *Pediculati*, which live in the mud, or in *Protopterus*, which for a part of the year is completely buried in mud (fig. 19, p. 44).

(6) *Atrophy of muscles.*—Cessation of use is also the cause of the degeneration of the flexor and extensor muscles of the fore-limb in Cetacea, and of the imperfection of the finger joint articulations in Cetacea and Sirenia (fig. 79). In the latter cases the surfaces of the articulation which make flexor and extensor movements easy disappear. When a limb becomes a paddle, it is necessary that it should be flexible, but that the articulations should be immobile.

(7) *Atrophy of the tail in man.*—The caudal region of the human vertical column is composed of four or five very degenerate vertebrae. The whole of this organ is degenerate. When the tail is formed at an early stage of embryonic life, the vertebral column consists of thirty-eight vertebrae; the lesser number of vertebrae in the adult is due to reduction of the tail, which in man is quite useless. Later on in life a further instance of atrophy may occur in individuals. In old men the caudal vertebrae are frequently fused, and the whole region is smaller.

Fig. 79.—*Globiocephalus melas.* Right anterior fin showing absence of articular facets for the joints of the fingers. (After Flower.)

(8) *Degeneration of the hyoid apparatus in man and birds.*—This case shows a close correspon-

dence between atrophy and loss of function. The second arch becomes connected with the third, the parts of which are, in the adult, the styloid process, the stylo-hyoidean ligament, and the lesser horn of the hyoid bone. According to the weight of the tongue, the parts of the second arch become more or less developed. In man the suspensory apparatus of the hyoid bone is extremely simple, and it is still more reduced in birds. "The tongue, reduced to a minute cartilage, no longer requires the support of a bony base so that the hyoidean apparatus might almost be removed from the anatomy of a bird. It is present, but in a rudimentary condition" (Geoffroy-Saint-Hilaire).

In the horse, which has a heavy tongue, the second arch is strong and completely bony.

In fish the hyoidean system is still more important, although in them this is associated not with any importance of the tongue, but with the branchial apparatus. The parts of the second arch are very strong, as they form a fulcrum against which the branchial system works, but its main parts are recognizable. "The hyoidean apparatus is the same in all vertebrates; its functions are at a maximum in fish, and at a minimum in birds, while in mammals the condition is intermediate" (Geoffroy-Saint-Hilaire).

II. TRANSFERENCE OF FUNCTION.—Among plants and animals there are many instances of organs well developed in younger stages of life, but which become rudimentary in later stages on account of their functions being assumed by other organs.

(1) *Atrophy of the tail in Batrachia Anura, and of the larval gills in some insects.*—The tadpole of the frog has a well-developed tail which acts as the organ of locomotion; the adult animal moves by its limbs, and the tail, useless in the adult, has been removed by phagocytosis.

The aquatic larvæ of many terrestrial insects possess tracheal gills, that is to say, membranous expansions of the skin, within which tracheæ ramify. In the adult insect respiration is conducted by normal tracheæ communicating with the air, and the larval organs of respiration atrophy.

In addition to such cases of atrophy occurring normally in the life-history of individuals, there are known many cases where the organs of a species have disappeared on account of the transference of their functions.

(2) *Disappearance of limbs.*—Vertebrates which move by general undulations of the body have lost their limbs for this reason. Such cases are Slow-worms, Amphisbæna, Snakes, Eels and Cæcilians.

Many parasitic creatures have similarly lost their organs of progression, as they depend upon their host for movement from place to place.

Sacculina, a parasite on the carapace of crabs, has completely lost its organs of locomotion. Moreover, as it takes its food by processes passing into the tissues of the crab, its digestive canal has been lost.

(3) *Atrophy of the leaf.*—In many plants the leaves have disappeared, their function having been assumed by some other part of the plant, as, for instance, by the phyllodes in *Acacia* and in *Phyllanthus* (fig. 84).

(4) *Atrophy of the protonema in mosses, and of the leaves in some xerophilous plants.*—At germination, mosses produce a much branched filamentous structure which serves as the organ of nutrition, and is termed the *protonema*. Later on this gives rise to buds which develop into the normal leafy shoots of the plants. As soon as the leaves are large enough to manufacture food for the plant, the *protonema* begins to degenerate, and disappears completely, except in a few rare forms (Ephemeraceæ) where the leafy shoots remain very small.

In *Muehlenbeckia platyclados* (fig. 80), which has become adapted to arid regions, and in consequence has the surface from which evaporation may take place much reduced, the branches which do not bear assimilating leaves are rounded at their bases, but higher up flatten into broad blades. These blades contain chlorophyll and fulfil the functions of leaves, these latter being present only as minute scales.

Similar phenomena occur in many Papilionaceous plants belonging to the genera *Genista* (broom), *Spartium*, *Alhagi*, etc. The leaves disappear, and the stems assume their functions.

(5) *The reduction or disappearance of the calyx.*—The disappearance of the calyx in many composite flowers and of roots in epiphytes are instances of species losing organs because of the transference of function to other organs.

In the simplest Compositæ, the fruits (*achenes*) are disseminated by the wind, the calyx usually being modified into a feathery tuft. In other Compositæ, although wind dispersal still occurs, the feathery calyx is lost, its purpose being fulfilled by wings on the sides of the fruit (*Florestinia pedata*), or it is united to a large membranous scale developed from the receptacle (*Dahlia*), or the sterile florets may be turned into wings (*Lindheimeria texana*). In other cases

Fig. 80.—Branch of *Muehlenbeckia platyclados*.

the fruits are dispersed, not by the wind, but by animals which eat them (*Clibadium asperum*), or they adhere to the fur of animals by hooks formed on the achenes (*Calendula*), or by hooks formed from the involucre of bracts (*Lappa*). In all these cases, and the list might have been made longer, the function of the calyx has been reduced, or has disappeared on account of the transference of its function.

(6) *Atrophy of roots.*—Roots in most plants perform two functions: they fix the plant in the soil, and, chiefly by means of the delicate hairs on their youngest parts, they absorb water and dissolved mineral substances. Sometimes, however, instead of having root-hairs, the rootlets enter into a kind of partnership with a fungus, which lives in their tissues, and absorbs by its processes the necessary food materials from the soil. The pine and beech are examples of this.

In some orchids (*Corallorhyza, Myrmechis*) the fungi lodge in the subterranean part of the plant, and the branches, having no function, disappear.

In most of the epiphytic Bromeliaceæ the roots are useful only to anchor the plants; the absorption of water takes place entirely through new organs developed from the leaves. The roots are few in number and small, and after the plant has obtained a firm position their growth almost ceases. In another species (*Tillandsia usneoïdes*) of the same family the roots have completely disappeared.[1] This plant, called by the Brazilians, "the plant of the air," fixes itself to branches of trees by its leaves, and is easily transported by the wind. Its

[1] For further details concerning the Bromeliaceæ see Schimper, *Die epiphytische Vegetation Amerikas*, in *Schimper's Botanische Mittheilungen aus den Tropen.* Jena, 1891.

absorption takes place entirely through hairs developed upon the leaves.

In addition to such examples of plants which have lost their roots on account of the functions of the roots being assumed by other organs, there are also plants in which the roots represent the sole

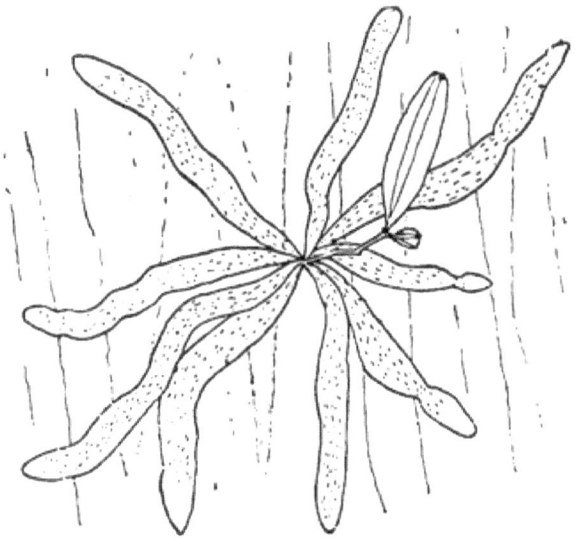

FIG. 81.—*Tæniophyllum Zollingeri* with branches adhering to bark. The plant bears a flower and a bud.

vegetative organs. In *Tæniophyllium Zollingeri* (fig. 81), an epiphytic orchid, the leaves are reduced to minute scales and are of no importance in the nutrition of the plant, that function being transferred to the roots which in the form of flat green ribbons, apply themselves closely to the bark of trees. In this case the roots have assumed the functions of leaves, while in *Tillandsia usneoïdes* the leaves have taken on the functions of roots. .

The Podostemaceæ are aquatic plants which live in warm regions attached to rocks in cascades. The stems and leaves are completely absent, the flowers even being produced on the roots. Some of the roots become closely attached to the stones. Others which are green and ribbon-like, float in the stream, and serve for assimilation.

§ 3. *Atrophy from lack of nutrition.*

In some cases of degeneration, the organs do not lose their functions, but become reduced, merely because adjacent parts rob them of their nourishment. This kind of lack of nutrition which results in local atrophy, must be distinguished from the general limitation of food-supply which is the ultimate cause of all degeneration. The possible amount of food within the reach of any organism is limited, but besides this, the share of the absorbed food obtained by any particular organ or part of an organ may be limited with a resulting atrophy of that organ or part. Atrophy of this kind may be accidental or normal.

1. *Parasitic castration.*—A good instance of accidental atrophy of this kind is found in *Melandryum album* where the ravages of a fungus *Ustilago antheridarum* may produce parasitic castration.[1]

[1] See Giard, *La Castration parasitaire* in the *Bull. Scient. de la France et de la Belgique.* 1887, 1888, 1889.

Under the influence of this parasitic fungus which makes its way to the anthers, the stamens of the female flowers assume the form of those in male flowers and in consequence the pistils of these flowers abort from defect in nutrition.

2. *Severe or prolonged compression of a limb.*—A long continued or forcible compression of a limb results in atrophy of its extremity on account of lack of nutrition from the compression of the blood-vessels.

3. *Atrophy of the genitalia in neuter bees.*—Lack of nutrition is also the cause of the arrested development of the genitalia normal in neuter bees. The neuters of bees and of some of their allies are females in a state of arrested development. In wasps, humble-bees, and hive-bees, it sometimes occurs that the genitalia of these forms develop sufficiently to be functional, thus resulting in the appearance of small females. In most honeycombs two kinds of cells are formed: in the smaller and more numerous cells are placed the larvæ destined to become neuters, in the larger and less numerous those destined to become queens or perfect females. The food of the two sets of larvæ is different; those in the larger cells are given "royal food" a more nutritious substance. When some of the royal food by an accident gets into a worker cell the sexual organs of that larva are developed so that a small female is formed. In this way as many females as may be desired can be produced,

and when a hive has lost its queen, the worker bees produce another[1] (Lacordaire).

Plants afford many instances of degeneration due to defect of nutrition.

4. *Atrophy of the superior flowers in Carex.*—In the tall spikes of Carex, it frequently happens that the flowers towards the summit are rudimentary, and authorities are agreed in regarding this condition as the result of defect in nutrition.

5. *Atrophy of pistils and stamens.*—Cases of atrophy of the stamens or pistils normal in species may be given.

In *Fritillaria persica* the flowers are disposed in bunches. The lower flowers possess six perianth members, two cycles of three stamens each, and a pistil. In the median flowers the pistil is smaller, and rarely capable of being fertilized. In the superior flowers the degeneration is complete, the pistil hardly being formed. It might be shown

---

[1] The transference of a function is not invariably accompanied by degeneration. Thus, in the functional development of an individual's nervous system voluntary acts which have been repeated frequently become reflex actions, and have their seat in a different region of the nervous system—as, for instance, walking and acquired professional dexterities. It has been sought to explain the development of the instincts of species in this way by supposing that frequently repeated voluntary acts have become inherited reflexes. As pathological degeneration in man frequently affects the higher regions of the brain, reflexes and instincts may persist after loss of voluntary action. In the cases of transference of nervous functions to lower centres the higher centres do not degenerate but remain able to acquire new voluntary functions.

ATROPHY OF ORGANS FROM LACK OF NUTRITION 277

that this degeneration is due to lack of nutrition, by removing the inferior flowers from a young bunch, but we are unaware that the experiment has been made.

In *Viburnum tomentosum* (fig. 82) the flowers are arranged in a compound corymb and occur in two forms. The central flowers have a small corolla, five stamens, and a well-developed pistil

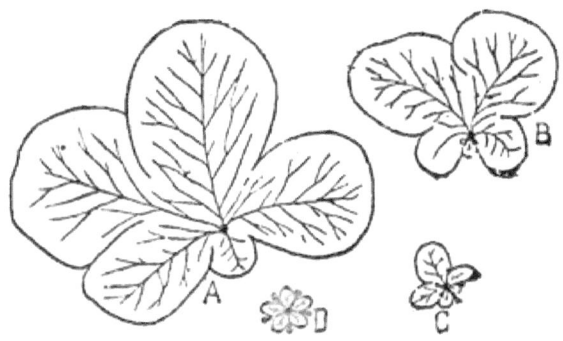

FIG. 82.—Flowers from the same inflorescence of *Viburnum tomentosum*.
A, sterile flower with rudimentary pistil and no stamens.—B, sterile flower with very rudimentary pistil and stamens.—C, flower with two fertile stamens; the other three stamens and the pistil are rudimentary —D, fertile hermaphrodite flower.

(fig. 82, D). The eight or nine peripheral flowers of each inflorescence have the corolla much larger, but the stamens are absent and the pistil is rudimentary (fig. 82, A). The petals turned towards the outer side are much the largest, and it is not rare for the petals turned towards the inner side to be hardly larger than those of the central flowers. What is important to notice, as it bears out the relation between degeneration and lack of nourishment, is that a smaller petal is always associated with a larger stamen (fig. 82, B, C).

In *Viburnum opulus*, the wild guelder-rose, there are also two kinds of flowers, the central flowers which are sexual and hermaphrodite, and the peripheral flowers which are sterile and possess a very large corolla. The five petals are, however, equal in size, and there are no forms transitional between the sterile and sexual flowers. The cultivated guelder-rose is a variety of the wild form in which all the flowers are sterile and possess a large corolla: the plant has completely lost the power of sexual reproduction.

§ 4. *Atrophy without apparent cause.*

In many cases it is impossible to determine the cause of atrophy. Why, for instance, has half the inflorescence disappeared from a unilateral cyme? Why have some composites like *Artemisia* or Rosaceæ like *Poterium* lost their habit of being fertilized by the agency of insects, and become anemophilous without conspicuous perianth? Nor can we explain why many Myriapods are blind, although they live in association with species possessing eyes and in conditions where vision apparently would be useful. Semper discovered, in brackish water in the Philippines, Crustacea (*Cymothoë*) which were completely blind, although they lived in light. The cause of such atrophies is unknown.

Moreover, we know that organs may atrophy through correlation with other degenerating organs, or because the whole organism is degenerating although there are no obvious anatomical bonds present between the related parts. As instances of such atrophies we may mention degeneration of the brain resulting on degeneration of the supra-renal capsules; or of the condition of myxodœma resulting from mechanical or functional disturbance of the thyroid gland. When the essential sexual organs of the male or female are congenitally or accidentally absent, the secondary sexual characters —the beard, voice, hair, and whole male or female aspect—of the body degenerate. When one eye becomes injured or diseased, a frequent consequence is sympathetic degeneration of the undamaged eye.

To these cases of correlative atrophy we may add the cases of leaves on many plants with drooping or horizontal branches. On these, the leaves on the upper aspect are frequently rudimentary. The appearance is most striking where the leaves are opposite the pairs being at right angles to one another. In a branch of *Acer campestre* which is upright, all the leaves are of the same size. In a drooping branch the leaves turned towards the sky are small. In plants belonging to very different families (Acanthaceæ, Melastomaceæ Urticaceæ, etc.) oblique branches exist, and the failure in development of the dorsal upper leaves is invariable. In *Procris laevigata* (fig. 83) the reduction of these leaves has gone very far, but in *Klugia notoniana* the leaves in the corresponding

Fig. 83.—Branch of *Procris laevigata*.
F′f′, F″f″, F‴f‴. The successive pairs of leaves in the axil of a large leaf F′ is a female inflorescence; in the axil of a small leaf f″ is a male inflorescence. (The original drawing was made at Buitenzorg in Java by Mas Kromohardjo, a Malay draughtsman.)

position are absent. In another representative of the same family (*Streptocarpus monophylleia*), the degeneration has gone still further; all the leaves are absent and a single greatly enlarged cotyledon is the sole assimilating organ.[1]

It is evident that in these oblique and horizontal branches, the leaves directed vertically towards the sky are in the most unfavourable position for assimilation, and that, in addition, their presence would shade the leaves lying under them. In these species the more or less complete reduction is an inherited fixation of a sacrifice of these particular leaves for the benefit of the whole organism.

CHAPTER II

ATROPHY OF INSTITUTIONS

THE causes of atrophy in institutions are more or less analogous to those which bring about atrophy of organs. First, there is *atrophy from want of use*, when function either becomes useless or is transferred to another institution. Atrophy from lack of resources corresponds precisely with atrophy from lack of nutrition. There is nothing, however, in the atrophy of institutions quite analogous to that which, in organisms, results from lack of space. But if the development of an institution cannot be actually impeded by the co-

[1] See the figures of plants of this family in Fritsch, *Gesneriaceæ*. (*Engler und Prantl's natürlichen Pflanzenfamilien*. Leipzig, 1891.)

existence of another institution, certain instances of atrophy do occur which cannot be said to arise from either want of use or lack of resources. The institution degenerates merely from hindrance offered to the exercise of its functions. Thus, for instance, the laws of exclusion voted in Germany against socialists brought about the decadence, and, ultimately, the downfall of a number of professional and trade associations which cannot be said to have been actually suppressed by law. It is the same with the corporations of Western Flanders; these have survived the revolutionary laws, but are no longer legally recognized, a state of things leading to many difficulties; the properties of several of them have been confiscated, as having no owners, and allotted to benevolent institutions. This want of legal recognition must inevitably lead in the near future to the entire disappearance of these vestiges of the corporative system.

Cases such as these, however, must be regarded as exceptional, and it may be said that, on the broad average, institutions atrophy either from want of use, or from lack of resources.

§ 1. *Atrophy from want of use.*

I. FUNCTIONAL INUTILITY.

(1) *Offices in connection with the Port of Bruges.*—The disorganization of the train service in a besieged city has already been alluded to as

furnishing an example of degeneration by suspension of function.

Another incidence of accidental degeneration is furnished by the office of wharf-porter, which was formerly exercised in the port of Bruges. When Bruges, consequent upon the blocking up of the Zwyn, ceased to be a sea-port town, the wharf-porters who were formerly employed to carry grain, lime, and coal, etc., were no longer required and so abandoned their calling.

(2) *The forest-courts of England.*—As an instance of normal degeneration due to the transformation of an agricultural country into a commercial country, take the old English forest-courts. In the middle ages there still existed in England great tracts of forest land which were Crown property, and subjected to special legislation conducted by three separate courts of justice : (*a*) The Court of Attachment[1] which instituted proceedings ; (*b*) the Court of Swainmote[2] before which the culprits were tried, and the Judge's Court presided over by the Lord Chief Justice, who pronounced sentence, and from whose decision there was no appeal. These courts have lost all importance since the seventeenth century, and the forest laws are now only functional

---

[1] When a forest law was infringed, it was the duty of the forester to "attach" the culprit—*i.e.* constrain him to appear either by seizing his person or his goods. These attachments were then submitted to the Court of Attachment.

[2] The judges were called *verderers;* the jury was composed of foresters of the *reeve,* and of four men out of each forest hundred.

with regard to the Forest of Dean and the New Forest. The office of Forest Judge has disappeared, and but few vestiges remain of the courts of attachment and Swainmote. Those persons who are connected with either the Forest of Dean or the New Forest, meet together still to transact business in connection with them, but these meetings are of no great importance.

II. Transference of Function.

Instances abound of the transference of a function from one institution to another; but as a rule, when this change is effected, it is attended by the suppression of the old institution. Sometimes, however, this is not the case, and the institution merely atrophies without having been suppressed.

(1) *Republican institutions under the Roman Empire.*—As instances of this form of degeneration take the institutions of the Roman Republic, after the establishment of the Empire—or the decadence of the functions formerly discharged by the Privy Council in England, the political functions of which have been transferred to Parliament, and the judicial functions to the Court of Common Pleas and the Court of Exchequer.[1]

(2) *Special jurisdiction in England.*—It is the

---

[1] De Franqueville, *Le Gouvernement et le Parlement brittanique*, i., p. 431.

same with special jurisdiction, such as the Ecclesiastical Court, and the University Courts; they have completely degenerated, their functions having been transferred to the jurisdiction of the Common Law Courts. In England, however, special jurisdiction has not wholly ceased to be functional, for there still exist, side by side with the modern Courts of Justice, a few local Courts and other exceptional forms of jurisdiction which are still maintained in support of certain ancient acquired rights and traditions.[1] By far the greater part of these special jurisdictions have, however, fallen into disuse, owing to the creation of the modern Courts of Justice. This happened, for instance, with the following institutions :—

(*a*) *The Local Courts of Feudal origin.*—These Courts have decreased both in number and importance since the close of the thirteenth century; the only vestiges now remaining of them are the Court-leets of certain manors. Sir James Stephen makes special mention of the Court-leet of the Manor of Savoy, which extended from near the old city gate of London (Temple Bar) up to Cecil Street. Some of the old functions of this Court are maintained in the present day.

(*b*) *Country Town Courts.*—Twenty-seven local courts are mentioned in modern judicial statistics before six of which no case has been heard for twelve years. On the application of a litigant,

---

[1] Idem, *Le système judiciaire de la Grande-Bretagne*, i., p. 216.

however, any one of these would resume its functions.[1]

(c) *The Court of Stannaries*, the vice-warden of which is appointed by the Prince of Wales. This Court has greatly degenerated in importance, and only a few minor cases are now heard before it.

(d) *The Court of Piepowder*.—This was a Court of summary jurisdiction, dealing promptly with disputes arising during fairs and markets. It has almost disappeared from having fallen into disuse. Practically only one example remains; that is at Bristol, and is becoming merged in another local court called the Tolzey Court.

(e) *The Husting's Court* (or folkmote, scirmote, a kind of County Court of the city of London).— The old *Husting's Court* has completely lost all its former attributes, in favour of the Court of the Lord Mayor and Sheriffs. Up to 1860, however, it reserved to itself the right of jurisdiction with regard to matters relating to landed property in the city; the only cases which come before this Court nowadays are those of replevy.[2]

Finally, there is the House of Lords which

---

[1] According to jurisprudence, the fact that a Court still existing by virtue of Royal Charter, has ceased to be functional for two hundred years, is no reason against a citizen having his case tried before it, if he so chooses, even though the town authorities declare the funds to be insufficient for payment of the judge.

(*Case of Rex v. Mayor of Wells, Dowling Practise Cases*, p. 562), mentioned by Franqueville.

[2] Franqueville, *Système jud.*, i., pp. 235, and following.

formerly held full powers of jurisdiction with regard to matters relating to members of the peerage; this jurisdiction is now limited to cases of high treason or crimes committed by peers to the exclusion of mere acts of misdemeanour which are tried before the ordinary Courts.

§ 2. *Atrophy from lack of resources.*

The instances we have mentioned are those of institutions decaying because they had become useless, and their resources were transferred to other objects. The lack of resources was the result of the lack of function. Sometimes, however, it is the cause and not the result, in which case atrophy may be due either to an abnormal development, such as the hypertrophy of another institution, or to poverty ensuing on the general decline of society in general. Of course both factors may act simultaneously.

1. *Local administration at the close of the Roman Empire.*—The decline of local government at the close of the Roman Empire is an instance of atrophy ensuing from the ultra-development of another institution.

As the demands of the central powers grew more and more excessive, the fiscal rates had to be augmented in order to meet them, and the *curiales*, which consisted of members of the City Council who were made responsible for the payment of taxes, finally found it was quite impossible to meet their engagements, and made every effort to leave the curia.[1]

2. *The degeneration of Societies in all their parts.* —A number of instances might be mentioned of general social degeneration bringing about the atrophy of some one or other institution in particular.

Besides giving classical examples, such as the Romans, Peruvians and Astecs, V. Lilienfeld mentions the decline of the Negro kingdoms which existed during the sixteenth and seventeenth centuries in Southern and Western Africa, and which are merely represented nowadays by wretched little tribes.[2]

There are, according to Waitz, at some distance from Carimango (the equatorial Republic) some people of pure Spanish blood who have fallen back into absolute barbarism. Their language is deformed past recognition, and their manners and customs exhibit no traces of their former condition.[3]

Space precludes us from dwelling further upon the various causes—often complex and obscure—which bring about the downfall of societies, suffice it to say that they are connected with

[1] Lavisse and Rambeaud, *Histoire générale*, I., ch. i., pp. 14 and following.
[2] Von Lilienfeld, *Gedanke über de Sozialwissenschaft der Zukunft*, ii., p. 241.
[3] Waitz, *Anthropologie der Naturvolker*, 1. B., p. 369.

territory or with population, the two factors of social evolution.

Either the physical surroundings of institutions undergo unfavourable transformations, or else the population itself degenerates.

(1) The almost complete disappearance of the great family communities (*zadrugas*) of Montenegro is a characteristic instance of atrophy from lack of resources caused by the impoverishment of physical surroundings. The persistent cutting down of trees in the Black Mountain has had a disastrous effect on the water supply, and consequently upon the fertility of the ground. Most of the *zadrugas*, having found it impossible to continue their existence in common upon the same territory, have split up into small families (*inokosnas*). These latter represent, in a reduced state, the old family system from which they have sprung. Bogisic has shown that these in no way resemble our modern families, but are to be regarded, from the judicial point of view, as reduced family communities, each comprised of only a few persons.

(2) Other cases occur where the atrophy of an institution—of an artistic or scientific society, for instance—is brought about by the degeneration of a population which ceases to be interested in the society and no longer contributes to its support. A large number of cases of this kind might easily be mentioned, especially as occurring during the period of the Byzantine Empire, but it is difficult

to account for the sudden degeneration of a people where the physical surroundings had recently undergone no special modifications, and when there had been no sudden and violent check upon social development. According to Lapouge and other sociologists of the Darwinian school, this social degeneration was merely the outcome of hereditary influences. The destiny of a nation is dependent upon the quality of the elements of which it is composed and by which it is directed. If a nation is rich in energetic and intelligent qualities, the greatest of disasters can only have a transitory and limited influence. When the contrary is the case, the same circumstances may produce an arrest in development or a complete decline and fall. Up to the present time, and especially in antiquity and the middle ages, these favourable qualities were generally supplied by a dominating minority establishing itself in a conquered country. In the common course of evolution, these superior elements, which are indispensable to social progress, are eventually eliminated. The inferior elements regain greater power, and each step of their progress is attended by a backward step towards barbarism. Although, at first sight, this seems contrary to the Darwinian theory, it is strictly in accordance with it. The superior individuals are relatively inferior when their chances of success or of posterity diminish. The superior individuals may not only be swamped by a diminution in their birth-rate, but in some cases there may be a direct elimination of them.[1]

The tendency of decadence is always towards the degenerative and eliminatory selection of superior elements.

It may be said in conclusion that there are constant calls upon the capital and labour of a society from its various institutions, and the consequence is that, the resources not being unlimited, a regular struggle for existence goes on amongst the institutions. In the course of this struggle, the decline of an institution may be brought about in two different ways. It either begins to degenerate from lack of sufficient means of support, or degeneration sets in consequent upon the institution having ceased to be functional by inutility, by transference of function to another institution, or by obstacles placed in the way of exercising that function. In either case the institution disappears. It is only in exceptional cases, which will be alluded to further on, that existence is still maintained.

[1] G. de Lapouge, *La Vie et la Mort des nations* (*Révue int. de Sociologie*, 1894, pp. 421 and following). Several terms used in this treatise were borrowed from the above article.

See also Hovelacque and Hervé, *Précis d'Anthropologie*, p. 189: "War, in its double consequence of the elimination of the strong and the survival of the weak, is for the more civilized races a powerful factor in the cause of degeneration and downfall."

PART II

THE CAUSES OF THE PERSISTENCE OF ORGANS OR INSTITUTIONS WITHOUT FUNCTION

CHAPTER I

SURVIVAL OF ORGANS

WE have shown how and why organs may become rudimentary and tend to disappear. In many cases the disappearance is complete; and the organ may not even reappear temporarily in the course of the individual development. This disappearance is, however, by no means universal. Even apart from the phenomena of recapitulation, rudimentary organs may persist in the adult, and sometimes, even although organs have ceased to be functional, they persist without degenerating. We have now to consider why in such cases degenerative evolution does not result in complete obliteration of such organs.

§ 1. *Unfunctional organs that are not rudimentary.*

ABSENCE OF VARIATIONS.—There are some plants such as *Ficaria*[1] *ranunculoïdes* and *Lysimachia*

[1] *Lysimachia Nummularia* occasionally produce seeds in some valleys of the Pyrenees, and Errera has shown us specimens grown from seeds coming from the shores of the lake of Quatre-Cantons.

Nummularia, the flowers of which hardly ever produce seeds. How is it that in such species flowers are still produced? The probable explanation of this anomaly is, that for the disappearance of flowers there would have to be produced individuals with this advantageous variation. It is the case, however, that the *Ficaria* and the *Lysimachia* reproduce most actively by asexual methods, and variations are extremely rare in cases of these modes of reproduction. The result is that these species having begun to form sterile flowers continue to produce them through simple lack of variation.

An analogous case is presented by *Elodea Canadensis*. This unisexual plant is represented in Europe by only female plants. These plants have multiplied asexually so luxuriantly that in Holland they began to choke up the canals, and it became necessary to make provision in the budget of that country for the extermination of the pest. The plants are, of course, able to multiply only asexually, as the female flowers cannot be fertilized, and these useless flowers have been maintained simply from the absence of variations.

Stratiotes aloïdes, a plant belonging to the same family as *Elodea*, is practically only represented by male individuals. Females are extremely rare, and none the less the male flowers are produced, although in the vast majority of cases they must be useless.

W. Burck has called attention to, without endeavouring to explain, other instances of the persistence of functionless organs. A large number of Anonaceæ bear flowers which do not open, and which are self-fertilizing (cleistogamous flowers). None the less, they have retained the corolla, the original purpose of which was to attract insects.[1]

Burck has called our attention to the circumstance that several species belonging to the same genus produce cleistogamous flowers, and that it is improbable that this condition has been acquired independently by these species. One would thus have to admit that the original type must be very remote, as it has given rise to descendants of specific distinction, and yet the useless corolla has persisted through the long series.

Parallel examples may be found among animals.

Machaerites is an insect which inhabits caves in North America. The females are quite blind; the males, on the other hand, have preserved, or seem to have preserved, well-developed eyes; but are these eyes real? An abyssal fish, *Ipnops*,[2] seems as if it had enormous eyes extending from the corner of the snout some distance along the neck, but these organs are not really eyes; they

---

[1] W. Burck, *Ueber Kleistogamie im weiteren Sinne und das Knight-Darwinsche Gesetz* (*Ann. Jard. Bot. Buitenzorg*, viii., p. 122, 1889).

[2] Dollo, *La Vie au sein des mers*, p. 242, Paris, Baillière, 1891.

are light-producing organs, and the fish are in reality blind. This may be the case also in the males of *Machaerites*. It may also be the case that the male has an opportunity for using eyes absent in the case of the female, the males sometimes leaving the caves, the females remaining within them. Something analogous to this occurs in the case of eels: the males remain always in the sea while the females rejoin them only for purposes of reproduction. Moreover, there is still a third hypothesis, that the male of *Machaerites* became an inhabitant of caves later than the female, and has not yet had time for the loss of its eyes.

§ 2. *Unfunctional organs which persist as rudiments.*

It is outside our purpose to discuss here the numerous cases of organs reduced through adaptation, such as the leaves reduced to serve as protectors of young buds (*Phyllocactus*, fig. 78), or the wings of the ostrich which, although much reduced, are supposed to assist the bird in running. The utility of such organs explains their persistence; we are concerned here with organs which, although useless, persist in a reduced form.

1. ABSENCE OF VARIATIONS.—In discussing organs which, although without function, have persisted in a complete state, we attributed the persistence to absence of variations. It is probable that the same cause operates in maintaining useless vestiges.

It is to be noticed, however, that the variability of vestiges is frequently considerable. The flowers of *Asparagus officinalis* are sometimes, although rarely, hermaphrodite. Usually they are unisexual and exhibit the organs of the other sex in every conceivable stage of degeneration. It is probable that the unisexual condition has been acquired recently, and that there has not yet been time for the operation of natural selection to cause the disappearance of the useless organs.[1]

[1] We have actual knowledge of the mode of disappearance of an organ in one case, and can see the part played by variability. In *Phyllanthus speciosus* (*Xylophylla arbuscula*) the adult plant has three kinds of branches: vertical branches, with rudimentary leaves; oblique branches which spring from the axils of these leaves and themselves bear in two rows very rudimentary leaves; and flat branches which are the chief organs of assimilation of the plant and which also bear rudimentary leaves in two rows. In the seedling, on the other hand, there are formed after the two cotyledons, one or two completely developed assimilating leaves upon an upright stem (fig. 84, A).

The flat branches grow from the axils of these, and bear, unlike the flat branches of the adult, assimilating leaves; higher up the vertical stem bears only rudimentary leaves with flat branches in their axils.

It may happen, however, that the seedlings bear, directly after their cotyledons, rudimentary leaves (fig. 84, B) and in this case the reduction of the leaves is present not only in the leaves borne upon the vertical stem but in those borne on the flat branches. These latter bear only a small number of assimilating leaves. Thus, we have in this plant a remarkable example of incomplete recapitulation: the seedling preserves in a functional condition organs that are rudimentary in the adult, but the species furnishes instances where these leaves cease to be functional even in the seedling.

2. Insignificance of the Rudimentary Organ.—

It frequently happens that rudimentary organs are preserved simply on account of their insignificance: the absence of organs so small would not be an advantage to the plant sufficiently great to be laid hold of by natural selection.

Many species of *Tropaeolum* bear leaves without

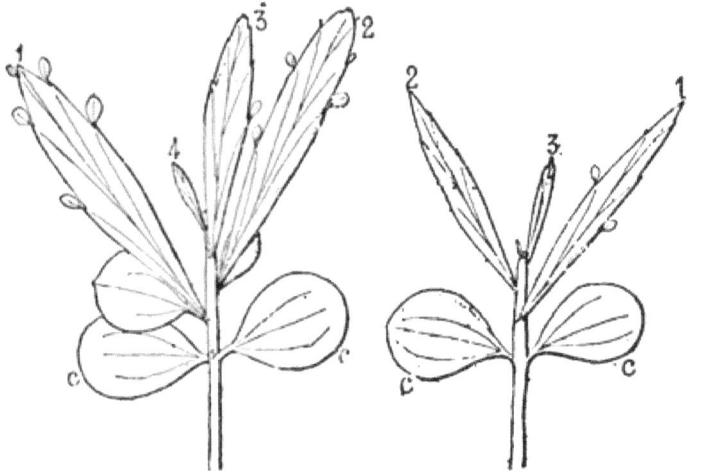

Fig. 84.—Two seedlings of *Phyllanthus speciosus*.
1, 2, 3, 4, successive assimilating branches; *c*, cotyledons.

stipules. In *Tropaeolum majus* there are stipules only in the case of the first two leaves of the seedling, and the position of these stipules is very variable. Sometimes both stipules are at the base of the petiole; sometimes only one is present; sometimes both are several millimetres from the base—a position in which functional stipules never occur.

In the same way may be explained the persistence of accessory rudiments of enamel organs in the development of teeth. Besides the rudiments of the enamel organs for the milk teeth and permanent teeth, there are additional organs present in a very variable condition and number, nearer the external surface. They are, however, very generally present, and are exceedingly similar to the youngest stages of the normal enamel organs. Kollmann and Gegenbaur believe that they are abortive rudiments surviving from an ancestral condition in which teeth were more numerous.

CHAPTER II

THE SURVIVAL OF INSTITUTIONS

We have seen that an institution which ceases to be functional without dissolution—either voluntary or enforced—ensuing, tends to atrophy and disappear, while its resources are appropriated by other institutions. It now remains to account for the fact that this atrophy does not generally end in total disappearance. The two following hypotheses may be made:—

1. The structure of a non-functional institution may remain intact.
2. The institution may survive, but in a rudimentary condition.

§ 1. *The integral persistence of an institution.*

A non-functional institution may survive and retain its structure and resources in the following cases :—

1. By the intervention of some superior authority to prevent its suppression.
2. When, while ceasing to be functional, it continues to be useful, though indirectly so.
3. When its existence is maintained out of respect for old traditions.

We will take these three hypotheses in succession, but it must be borne in mind that when a non-functional institution is maintained out of respect to tradition, or by virtue of an indirect usefulness, it is always by the intervention of legal authority. This legal authority, however, amounts to an expression of the public will, whereas an institution may also be maintained by the exercise of personal influence on the part of some one person.

It sometimes happens, too, that a non-functional institution continues to survive because its suppression would entail important changes in other parts of the social organization.

1. MAINTENANCE BY COMPULSION.—A useless institution is frequently maintained by compulsion, when its conservation is advantageous to those connected with it, or even to other persons.[1]

The following are a few examples of this :—

---

[1] Spencer, *Principles of Sociology*, vol. iii.

(a) A long list might be made of all the sinecures, now quite useless, that some governments insist upon maintaining for the advantage of those occupying the posts. Such were certain offices in connection with the Court in former days or the *avoueries* of the end of the middle ages.

"Like the Fief system," says Errera, when writing about the Massuirs, "The *avouerie* afforded an effective protection—military as well as judicial—against the various dangers arising in a still barbaric age. But, in the course of the last centuries of the middle ages, the obligations of feudal chiefs and the condition of *avouerie* disappeared; the reasons were that relative security was attained; militia was established, and the army, under the command of the sovereign himself, became better disciplined; and that there arose the organization of the *justices scabinales*, of bailiwicks, and of superior courts of justice. However, although the ancient offices disappeared, the emoluments attached to them continued to be drawn.[1]

(b) It often happens that institutions which have ceased to be functional are yet maintained as being a source of profit not only to those in direct connection with them, but to a considerable number of other persons.

Before the Reform Bill of 1832, when large towns like Leeds, Birmingham, and Manchester were unrepresented in Parliament, the House of

[1] Errera, *Les Massuirs*, p. 75. Brussels, Weissenbruch, 1892.

Commons contained seventy members, nominated by thirty-five rotten boroughs in which there were no electors, and ninety members nominated by forty-six boroughs, containing less than fifty electors.

The borough of Old Sarum was a mere hillock belonging to Lord Canalford; Gatton and S. Michael had only seven electors; the borough of Dumwich had been long since submerged by the encroachment of the sea; Beeralston, belonging to Lord Beverley, consisted of one house, and Castlerising of two. In the county of Bute there were twenty-one electors, only one of whom was a resident and who nominated himself.

The preservation of this system of representation which had long ceased to be adequate, was eminently favourable to the few persons who benefited by it, and they vigorously resisted the passing of the Reform Bill.

(c) After the provincial states of Normandy and the Dauphiny had been suppressed, the state functionaries retained their titles and emoluments.

In the Dauphiny the representative institutions ceased to be functional in 1627, but at the close of the eighteenth century the Bishop of Grenoble continued to receive 6000 livres per annum as primate of the dominion. Two barons, delegates of the nobility, shared a similar salary, and the Syndic of the province and two secretaries received lesser emoluments from the province which continued to make these payments for services no longer rendered.[1]

In each of the above cases of survival it is plain that compulsion intervened—more or less directly—to secure the maintenance of non-functional institutions.

In the first cases referred to, the privileged persons concerned, took advantage of their influential positions to enforce the maintenance of their sinecures.

In the second case—that of the rotten boroughs—the institution was not only advantageous to the member himself but to the whole of his party, so that naturally its maintenance met with the cordial support of the latter.

In the third case, those in authority maintained part of the institution—that of the mere titles and emoluments—in order to suppress the remainder more easily. In this case it was not an institution which atrophied, but an institution which was caused to atrophy by compulsion.

2. INDIRECT USEFULNESS.—It sometimes happens that an institution, although ceasing to be functional, yet retains a certain usefulness. This is so in England, with the office of the Privy Seal. All the functions formerly discharged by the holder of this office have long since disappeared, but the post is reserved as a sinecure for persons who have

[1] Babeau, *Les assemblées des pays d'Etat sous l'ancien régime;* Réforme sociale, 1893, p. 704.

distinguished themselves in politics, but who from advanced age are unable to take an active part in public affairs. A great many sinecures are maintained for a similar purpose—that of furnishing practical though not nominal pensions to distinguished persons in art or science.

"It may well be," says Viollet, "that an institution, which at first sight seems to be a mere useless wreck, is really of immense service to society. The deep roots of a dead tree may continue to furnish support to a new structure."

It is by reason of this negative usefulness that —according to Bagehot—the English monarchy has been preserved, forming as it does a symbol of unity and coherence amidst the electoral changes of power. The English monarchy offers a characteristic instance of an institution which has lost nearly all its former functional importance, while nominally retaining its power. According to Bagehot, the Queen would now have to sign her own death warrant if condemned by both Houses of Parliament. The outer form, however, remains almost the same as in the days of absolute monarchy, when the sovereign took an active part in public affairs. At within the last few years, the Queen's signature was required to all official documents. It was only in 1862 that a law was passed deciding that for the future, promotions in the Army or Navy should be signed either by the Commander-in-Chief, or by the Secretary of

State. At that time, by dint of hard work, the Queen had signed all commissions up to 1858, and there remained 15,931 documents of this kind still unsigned.[1]

3. RESPECT FOR TRADITION.—The persistence of some institutions can only be accounted for by a lack of invention. Mommsen calls attention to a remarkable instance of this kind in the history of early Rome :

When a government by prætors—as consuls were first called—was substituted for a government by kings, the new system remained the same in idea though nominally different. The old idea of royal authority survived for a long period, and the prætors enjoyed all the old kingly powers, even those in plain contradiction to the temporary character of their office: the king could not be deposed, but neither could the prætor be constrained to depose himself; the king, when dying, nominated his successor himself, and this power remained to the prætor—although the system of election by the comitia had been introduced—for the prætor had the right of excluding whoever he chose from the list of candidates, and also of annulling the votes given to those who displeased him. It was only at a later period that a logical

---

[1] Bagehot, *The English Constitution*, pp. 57 and following; London, 1891.

De Franqueville, *Gouvernement et Parlement Britanniques*, i., p. 251.

and consistent conception of the consular authority came to be formed.[1]

In the same way the maintenance of the institution of sheriffs in modern England can only be attributed to an exaggerated respect for tradition.

In the commission which sat in 1888 to inquire into judicial organization, one of the commissioners expressed himself as follows : " I cannot see that sheriffs are of any use whatever, unless it be for show ; there is not one single function attached to the post which a sheriff can fulfil himself. I am a sheriff, but I do not know what a sheriff's duties are." The institution, however, remains intact. The sheriff is credited with the discharge of several functions, which are executed in his name and on his responsibility. All he does is to receive the judges, accompany them on circuit, and preside over parliamentary elections.[2]

The mode of nominating the sheriffs has remained unchanged since 1340. The Queen, by means of a traditional gold pin, is supposed to prick by chance in the list of candidates which is presented to her, the name of him upon whom she confers " the

---

[1] Ferrero, *Simbolii*, p. 53, Torino, 1893.

[2] "The two sheriffs of London, who are elected from among the liverymen of the various city companies, have not to go on circuit, but are supposed to attend at the Central Criminal Court. Their functions chiefly consist in escorting the Lord Mayor to all city ceremonies, and in attending numerous banquets, some of which are given at their expense." De Franqueville, *Système judiciaire de la Grand-Bretagne*, i., p. 611.

charge and keeping of the county." This ceremony is known as the "Pricking of Sheriffs."

§ 2. The Survival of Institutions in a Reduced State.

It has been shown that a number of institutions in a state of decline continue to be maintained, because they are still useful in spite of their reduced condition.

This is the case with the symbolic ceremonies which in former days accompanied the drawing up of solemn contracts.[1]

We now come to institutions which persist in a reduced condition, but which are not directly useful to any one. In this case their persistence may be attributed to one of two causes: either to respect for tradition or to the insignificance of the vestiges which remain.

---

[1] According to Viollet in *Histoire du droit civil fr.*, p. 607, the primitive assembly of the people still survives, though in a reduced condition, in the Roman *mancipatio*, and in Scandinavia in a solemn form of sale called the *scotatio*.

"I believe," he says, "that it was the primitive sale of German law a sale concluded and ratified in a popular assembly, that gave rise, in the middle ages, to the Scandinavian *scotatio*. So also in the case of the Roman *mancipatio* there has long been believed to exist the remnants of a popular assembly. The dumb witnesses in the *scotatio* appear to me the petrified representatives of the German tribe or village ; and, in the opinion of good judges, the witnesses in *mancipatio* are no other than symbolic statues of the five classes of the Roman people. However, this is mere hypothesis."

It is difficult, however, to distinguish between these two causes, for insignificant vestiges of institutions are especially numerous in very conservative environments, and, on the other hand, mere respect for tradition very rarely ensures the maintenance of either a harmful or an expensive institution.

1. INSIGNIFICANCE OF THE INSTITUTION.—Some of these insignificant institutions are reduced to mere vestiges, no longer functional, or of use to any one, others—representing traces of a former system—retain some local vitality, although not in keeping with the new conditions; these are of such small importance that their very insignificance ensures their survival.

(*a*) The two French laws passed in 1835 and 1849 relating to entail have never become functional at La Martinique. The old law of entail is still active there, and this state of things is allowed to continue without authoritative interference.[1]

(*b*) In England, the sovereign, up to the close of the sixteenth century, reserved to himself the right of presiding over the Courts of Justice, and of pronouncing sentence. Since the Revolution, no sovereign has essayed to render justice personally, and any attempt of the kind would be regarded as unconstitutional nowadays. Vestiges of the old system remain, however, in the formula of certain legal proceedings, such as the serving of warrants,

[1] Viollet, *Hist. du dr. civ. fr.*, p. 883.

in which an order is given to appear before the Queen herself, and it has never been suggested that the old formula should be altered.[1]

The following are a few examples of local survivals which owe their existence to their insignificance :—

(a) Although Cambray has belonged to France since the close of the seventeenth century, its diocese includes a small part of Belgium; the ancient religious organization has in this instance survived political changes.[2]

(b) It is not generally known that with certain properties situated in Artois—consisting chiefly of marsh land—the law of primogeniture still holds good.[3]

(c) Public attention has recently been directed to a very curious survival of the political connections which existed during the middle ages between Béarn and Spain. Every year, on the 13th of July, the inhabitants of the French valley of Baretous in the Pyrenees, solemnly pay a tribute—for the

[1] De Franqueville, *Système judiciaire de la Grande-Bretagne*, p. 23.

[2] Viollet, *Hist. du. dr. civ. fr.*, p. 882.

[3] Errera, *Les Masuirs*, p. 290. Decision of the Council of 25th February 1779, as regards the castle-wards of Lille, Douai, and Orchies. "Those portions of land which fall or have fallen to each inhabitant as the result of division, shall be inalienable; no person shall possess two portions. The eldest male of each family, or in default of the eldest male, the eldest female, shall alone inherit the said portions of land."

maintenance of peace—to the inhabitants of the Spanish valley of Roncal.[1]

It is plain that this custom, which has gone on so long as to pass unnoticed, is too inconsistent with the present relations between France and

[1] The ceremony commences at nine o'clock in accordance with instructions laid down in a document which, according to the mention made of it in the Procès-verbal, dates from 1375. First, the French mayors don their scarves of office; next the Spanish mayors, advancing from a group of compatriots, proceed towards the frontier boundary line, accompanied by a peasant carrying a lance with a red flag tied to it—a symbol of justice—and draw up within six yards of the boundary. The French mayors follow suit, but the flag hoisted on the lance which precedes them, carried by a peasant, is white, as signifying their pacific intentions. The Mayor of Isaba then says to them, "Is it peace?" The French mayors reply in the affirmative, and, as a proof of their sincerity, couch their lance upon the mile-stone marking the boundary. The Spaniards then first plant their lance in French soil leaning against the stone, and afterwards place it so as to form a cross with the French lance. Next the Mayor of Arette places one hand upon the crossed lances, and the Mayor of Isaba does the same, and together they utter the formal declaration of peace, which all those present swear to observe. After the vow, the Mayor of Isaba cries three times, "Paz davans!" which means, "May peace continue."

Peace being thus declared, the Roncalais, in order to ratify their abandonment of hostilities, order the guards to remove their arms from the French side. The ceremony being over, it only remains to pay the blood-tax. This formerly consisted of three white mares, all exactly alike; but, owing to the great difficulty of matching them, three unblemished heifers were substituted, all of same colour and with the same markings. These three heifers cost about 600 francs, which is a large sum for that part of the world.

Spain to be continued much longer. "It is to be expected," says the paper from which this information was obtained, "that the French Foreign Minister will in the near future come to an understanding with the Spanish authorities to put an end to this iniquitous custom, and it is to be hoped that this year is the last occasion upon which a blood-tax will be paid by the valley of Baretous to the valley of Roncal."

2. *Respect for tradition.*—In his "Essays on Progress, Manners and Customs" (*Westminster Review*, 1854), Spencer points out the connection between respect for tradition and custom, and the conservatism of those in authority. He says that certain customs, which have elsewhere died out, survive in some departments of the government. The Secretary of State in ratifying acts passed in Parliament uses old Norman French,[1] and certain legal terms in old Norman French are still used. The wigs now worn by judges and barristers are identical with those seen in old portraits, while the "Beef-eaters" of the Tower of London wear the

[1] "For financial Acts the formula is: La Reyne remercie ses bons sujets, accepte leur bénévolence et ainsi le veult ; for general Acts : La Reyne le veult ; for private bills : Soit fait comme il est désiré ; for petitions : Soit droit fait comme il est désiré. The veto is announced : La Reyne s'avisera. Cromwell had changed these old forms ; he gave his consent to Bills in English ; the old custom was resumed at the Restoration, and the House of Commons in 1706 rejected a Bill passed by the Lords to abolish the French phraseology." De Franqueville, *Gouvernement et Parlement Britanniques*, vol. i., p. 279.

same costume as that once worn by the body-guard of Henry VII.[1]

Two similar examples may be added:—

(a) At the coronation of English sovereigns two gentlemen of the Privy Council, chosen "on account of their appearance," and created knights for the occasion, are appointed by the Lord Chamberlain to represent the Dukes of Aquitaine and Normandy.

(b) The First Officer of the Crown was formerly the Lord High Steward, which title having, in the course of time, become purely honorary, was hereditary in the family of the Earls of Leicester. The post is now in abeyance, but, on the coronation of a sovereign, or, on the occasion of a peer being placed upon his trial, this dignity is conferred upon some important person nominated solely for the occasion. It is not only in ceremonial—which, according to Viollet, is the museum of history—that reduced institutions, which are completely useless, are tenaciously maintained; it is the same with judicial and religious institutions. A few examples will suffice to show that this is so:—

> (a) In its primitive form, the Assembly of the People included the whole army, and was necessarily held in some large open space. The custom survived the necessity of the choice of some such spot, and up to the sixteenth century, whenever a new Emperor was proclaimed in Germany, it was the

---

[1] H. Spencer, *Morals, Science and Art*.

custom for the electors to proceed to some mountain for the purpose, probably because it was the custom in former days to hold a general meeting there, before the election.

In Iceland, the *althing*—consisting of two chambers now—was formerly one large assembly held in the open air upon the Logberg (the Mountain of the Law) near to the Lake of Thingvellir. In the Republic of Andora, criminal sentences are still pronounced with great solemnity from the Market-Place.[1]

(b) Among the Ossetes, where the family community still flourishes, the only inalienable and unsaleable property is not the real property, but certain personal possessions, such as the great cooking-pot with the chain by which it is hung over the fire. "At first," says Kowalewsky, "this may appear a strange custom, but it must be remembered that possessions of that kind were of equal importance to the Ossetic 'family' as were their tombs to the ancient Greeks and Romans, which explains how any infringement of the

---

[1] In Montenegro, it was the custom, till recently, for the Prince himself to render justice, sitting under a tree in front of his palace at Cettigne. This mode of jurisdiction, which was probably a survival of the old system of which we have been speaking, has not completely disappeared.

custom is regarded as an infamy by the Ossetes."[1]

We now come to cases of purely religious survival, which offer the strongest resistance to the inroads of change. Spencer instances the custom of circumcising with a knife made of flint, and the vestiges remaining in Catholic worship of former primitive religions. The Eucharist, as we have already pointed out, is reminiscent of real sacrifices, and the symbolic representation by a dove of the Holy Ghost is only a rudimentary form of zoolatry.[2]

In Belgium there are still traces of the old custom of sacrificing an animal upon the completion of a new building, with the idea that the animal's spirit will protect the edifice from harm,[3] and if the observer of the following facts is correct in his interpretation of them, there also remain in Belgium traces of the ancient sacrifices to the genius of the earth :—

"It is the custom, round about Florenville (in the Belgian Ardennes), to offer a sacrifice to the presiding genius of the road upon the construction of a new road or railway. It is usually a fowl, or a rabbit, or even a calf which is sacrificed. . . . In some parts of Luxembourg animals

---

[1] Kowalevsky, *Droit coutumier Ossétien*, p. 105.

[2] Spencer, *Principles of Sociology*.

[3] *Folklore*, Wallon, No. 1526, p. 115 (*Bulletin de Folklore*, ii., 177).

are also offered in sacrifice to the genius supposed to preside over a newly-purchased field, with a view to ensuring abundant crops.[1]

Remains, either undoubted[2] or only probable,[3] of phallic worship, are scattered throughout Europe.

In Brittany[4] and in Belgium,[5] for instance, strange old customs still exist showing that here,

---

[1] *Révue des traditions populaires*, 1893, p. 394.

[2] Th. Volkov, *Rites et usages nuptiaux en Ukraine* (*l'Anthropologie*, 1891, p. 167).

Only a short time ago it was the custom in Tver, on the day dedicated to Yarilo (the phallic God of Spring), for the parents of young daughters to send them to join in games similar to those of the ancient Slavs, with a view to their getting married.

[3] *Note sur un vestige du culte de la terre mère* (phallism) *en Provence*, by Bérenger-Féraud (*Révue d'Anthropologie*, 1888, p. 563).

"At Luc, in Provence, upon the 1st of May, which is a country holiday, the young girls proceeded to a place where two roads met. Here they assembled around an olive tree, and after each dance they struck the olive tree three times with their backs.

"This fête, a survival of the floral fêtes of the month of May which are still celebrated in Provence and Italy, continued to be held until quite recently, and appears to have been a lingering vestige of the ancient worship of creative Nature, Mother-earth — in short, of phallic worship.

"The three knocks given by the young girls to the tree trunk is a survival of the ancient virginal sacrifice to the phallic emblem. The original meaning was not quite lost, for the Provençals still realized, though vaguely, that the three knocks were somehow connected with the idea of marriage."

[4] "*Les Mégalithes de Locmariaquer et de Carnac, et les amours*, by Bonnemère (*Révue des traditions populaires*, 1894, p. 123). "In former days it was the custom for all the young women who wished to get married to climb (on the night of May 1), to the top of the great *menhir* where they lifted up their clothing that their bodies

as elsewhere, a belief in the influence of fetiches once prevailed, and particularly in the form of megalithic monuments relating to the fecundity of women.

The custom still prevalent in the African Congo, of driving a nail into a fetich, with the view of reminding it of a request, has not disappeared from Europe. The young men of Couvin (Namur) still stick pins into the wooden saints of the little chapels round about in order to draw a lucky number in the military lottery, and young girls in Brittany do the same with a view to getting

---

might come in direct contact with the stone, and then slid from the top to the bottom."

At Carnac, young girls wishing to marry undressed completely upon the same night, and proceeded to rub their abdomen against a special menhir. In that part of Brittany, where only French is spoken, similar customs have equally prevailed.

[b] *Fête de Notre-Dame de Ride-cul* (Jules Lemoine, in the journal *Le petit bleu*, of October 18, 1896).

Similar to the above Breton custom is that of sliding down the Rocher Ride-cul, which is situated near to Landelies, in the valley of the Sambre. Here, as in many other places, old customs have been Christianized, and a Christian chapel now stands close to the ancient shrine. Young people of both sexes used to seat themselves upon the top of the stone upon little fagots of boxwood gathered in the neighbourhood, and then slide down to the bottom.

According to the old saying an upset meant waiting; an embrace signified mutual affection; a collision, indifference; and an embrace followed by rolling over indicated matrimonial suitability. Similar customs prevailed at Trou-deux-Trous, situated near to the Rocher Ride-cul. These two megalithic temples disappeared about forty years ago, their materials have been used as fluxes in smelting works.

married,[1] and in Belgium,[2] as in Brittany,[3] if the image prove too hard, the reminding pin is stuck instead into a fissure, or into the door of the niche containing the image. Besides these mere vestiges of pre-historic customs and belief, by referring to certain illustrated documents bearing upon the subject, it would be easy to work out the connection between the worship of saints which is prevalent in Belgium in the present day, and the pagan worship of the ancient Celts and Germans.[4]

[1] *Bulletin de Folklore*, i., 250-251.

[2] *Ibidem*.

[3] The menhir of the Pierre-Frite in the valley of Lunain. In nearly every hole or fissure of this monument, a nail or pin has been stuck by the young people of that part in the belief that it will ensure them a speedy marriage. (*Révue des traditions populaires*, 1893, p. 448.)

[4] See for Saint Eloi, *Mélusine*, viii., 122-132; for Saint Martin, *Bulletin de Folklore*, i., 309-315; for Saint Hubert, Gaidez, *La Rage et Saint Hubert*.

PART III

RESUMÉ AND CONCLUSIONS

WHEN an institution or an organ ceases to be functional or in any way useful, it very soon disappears altogether. If, as happens in some exceptional cases, it persists, it is because neither of the chief factors in causing atrophy, variability or selection, have intervened.

Sometimes the vestiges are of too insignificant a nature to call for their removal by either artificial or natural selection, and sometimes their existence is ensured by the lack of variability, as in the case of the persistence of flowers in plants which multiply asexually. This absence of variation occurs equally in the social domain, especially in matters connected with religion, wherein ancient customs are credited with a divine origin. Religions may pass away, philosophies may be transformed, and old beliefs cease to prevail, but the remnants of old creeds, conveyed by popular tradition through the centuries, defy destruction by modern innovations.

The ancient winter festival, on which day the

dead were supposed to leave their graves and join the living in a feast around the family hearth, is still celebrated in the keeping of Christmas and in the various customary practices on the first two days of November.

The May-Day festivals—pagan festivals held in honour of vegetable and human fecundity—are still held in their early form round about Locmariaquer and in the village of Campine. Traces also remain in the picking and wearing of flowers on the 1st of May, and the same day is selected by the socialists for the celebration of their near approach to a life under freer and happier conditions.

This survival of festivals, customs and traditions, while the religions and civilizations which produced them have passed away, is the principal link which connects us with bygone generations.

"Their value lies," says Houzeau in his *Étude de la Nature*, "in the establishment of a chain between successive generations. The memory of an individual may be regarded as constituting his personality. Take from him the memory of his past, and he is left at a point in time wherein there is no stability and complete isolation. To be himself, a man requires not only his recollections, but a knowledge of his past habits and traditions. When a savage is removed from his fellows and transported to new surroundings in a distant country, he loses all knowledge of his former condition. Society itself, made up as it is of customs and prejudices, constitutes history. The mirror of the past is exhibited in the consciousness of the collective individual which is called a nation. What link shall we have with former generations if not a heritage of their ideas—*i.e.* of their discoveries and their mistakes? Nations, like individuals, are continually modifying this inherited legacy, but, like the individual, they cannot get away from it without breaking the thread which has made them themselves."

GENERAL CONCLUSIONS

I

All evolution is at once progressive and retrogressive.

All modifications of organs and institutions are attended by retrogression. This occurs equally in the modifications of organisms and of societies. All existing forms, whether organic or social, have undergone certain modifications, and, as a result, have lost some parts of their structure. This universality of degenerative evolution may be proved either by the comparative method, or by showing that all organisms contain rudimentary organs, and that all societies contain survivals.

II

Degenerative evolution follows no definite path, and can in no way be regarded as constituting a return to the primitive condition.

In some cases—when one cause of dissolution equally and simultaneously affects all the parts of an institution or an organism—the most complicated and delicate structures are the first to disappear; but it must not be taken as a general principle that the most complicated structures are necessarily the most recent, and that consequently degeneration always retraces the path of progress. Evolution is irreversible, and accordingly, with a few more or less obvious exceptions, we draw the following conclusions :—

1. That an institution or an organ which has once disappeared never reappears.
2. That an institution or organ once reduced to the condition of a vestige cannot be re-established and resume its former functions.
3. Neither can they assume fresh functions.

III

Degenerative evolution is brought about by a limitation in means of subsistence — either in nutriment, capital or labour. In biology the principal if not the sole agents in its accomplishment are the struggle for existence between the various organs, and the struggle for existence between the various organisms.

In sociology it is artificial selection which is the dominating agent, and natural selection plays only a secondary part.

The occasional causes of degenerative evolution are inutility of function, insufficiency of nutriment or resource, and (in biology only) lack of space.

An institution or an organ which has ceased to be functional, and has also ceased to be useful either directly or indirectly, continues to exist if neither variability or selection intervene.

www.ingramcontent.com/pod-product-compliance
Lightning Source LLC
Chambersburg PA
CBHW030731230426
43667CB00007B/678